MathWise
Linear Equations

Peter Wise

MATH TEACHER,
MONUMENT,
COLORADO

CONTRIBUTORS

David Wise

Katherine Wise

Cover Design by Elizabeth Novak

Dedicated to Silva Chang,

founder and director of the Colorado Math Circle.
She has taught, coached, and prepared the top math students in Colorado
for success in state and national math competitions for years.
http://www.coloradomath.org/

I want to express my appreciation to Katie Wise, for her
insights, opinions, and helpful ideas

...with additional thanks to talented student editors who offered
constructive opinions and suggestions to improve the book:

Keiran Berry, Maren Busath, Parker Johnson, Aleks Solano,
Ryan Fowler, Aubrey Huffman, Brenna Locke, Abby Monforton,
Blaise McCabe, and Ethan Twesme.

MathWise Linear Equations

Copyright © 2015, Peter Wise

MathWise Curriculum Press

First printing 2015

MathWise Linear Equations

TABLE OF CONTENTS

TABLE OF CONTENTS, CONTINUED

PREREQUISITES FOR STUDYING LINEAR EQUATIONS

The study of linear equations is a good sequel to a study of basic algebra. Solving linear equations reinforces algebra skills while adding a new topic: graphing lines. To be ready for linear equations students should also be comfortable with negative numbers, substitution, and the distributive property. If students haven't acquired the necessary skills, it may be advisable to have them become acquainted with these topics before attempting the study of linear equations.

ABOUT LINEAR EQUATIONS

The study of linear equations teaches students the basics of graphing. They learn about plotting points, making x/y tables, reading and interpreting tables of values, observing proportional relationships, and testing for functions.

A line plotted on a graph has two aspects: slope (angle) and y-intercept (height). The most common formula for a line is the Slope-Intercept equation. It is essential that students understand that the m and b values (slope and y-intercept) are fixed numbers, the solutions for the x and y pairs are infinite. Knowing this, students will be better equipped to represent x and y as variables when writing equations for lines.

Observing the slope of lines is a great way to visualize fractions. The key to doing this is to note that a slope of 1 is a 45° angle going up to the right. Fractions between 0 and 1 fall below the slope line of 1. Fractions and whole numbers from 1 to values approaching infinity fall between the 45° line and the y-axis. It is helpful for students to understand the reciprocal relationship between values on both sides of the slope-of-one line.

I recommend that students, especially at the beginning, use the "parentheses trick" — which provides a container in which to substitute variables (see page 17). This is helpful for students just starting to plot points with x/y tables.

This book emphasizes the Slope-Intercept formula, even when determining equations of lines. Linear equations have four pieces of information: x, y, m, and b. As with proportions, if you are missing one piece of information, the other three values will enable you to figure out the fourth value. When solving for missing values in the equation of a line, I instruct students to list all the values they know for these four pieces of information. It helps students to see what they know from the information given, and exactly what it is that they are trying to figure out.

The book concludes with advanced material relating to linear equations: systems of equations. Three methods of solving for x and y are presented: graphing, elimination, and substitution. These studies are introductory in nature and can be used for enrichment or pre-teaching.

This series of books is designed to be unique and to catch students' attention in special ways:

Tips and Tricks
Over the years, I have assembled a wide assortment of memory aids—my tips and tricks. Students have found these to be helpful and memorable, but they have also found that these pointers add interest and excitement to their math studies.

Answer Frames
A distinguishing characteristic of this series is the use of answer frames. These frames teach students to show work and solve problems in a step-by-step manner. Students develop good technique as they solve problems. Books that just have white space leave some students unprepared to tackle the problems and allow them to make less-than-ideal use of their workspace. Answer frames have been a great help to my students in developing proper math technique.

Speech Bubbles with Teacher Insights
Speech bubbles are used to provide guidance, point out insights, or give helpful hints as students are solving math problems. Students learn best by doing, and the instruction given in the speech bubbles is designed to (1) sharpen students' powers of observation, (2) increase number sense, and (3) instruct in digestible chunks.

Simplicity of Instruction
Concepts are explained clearly and simply. Every page (excluding review pages or quizzes) has a specific focus. Most pages have generous amounts of white space to keep students focused. Movement is from the simple to the increasingly complex.

Step-By-Step Procedures
Students learn best when given explicit, step-by-step instruction. When several steps are involved, they are numbered. This makes learning much more logical and memorable.

Depth and Complexity
Throughout the book there are challenge problems to stretch students' thinking. At your discretion, you can guide students through the more challenging problems (recommended) or, alternatively, you can have them skip these harder problems.

Informal Terms
This book often employs informal language like "top number" or "bottom number" to keep things simple and focused. Standard mathematical terminology, such as numerator and denominator, is used after the concepts are presented.

Logical-Sequential Instruction
Math problems are presented in a logical sequence, so that previous problems contribute to students' abilities to solve future problems. The order in which you present math problems is critical to promoting number sense.

Introduction to Graphing in 4 Quadrants

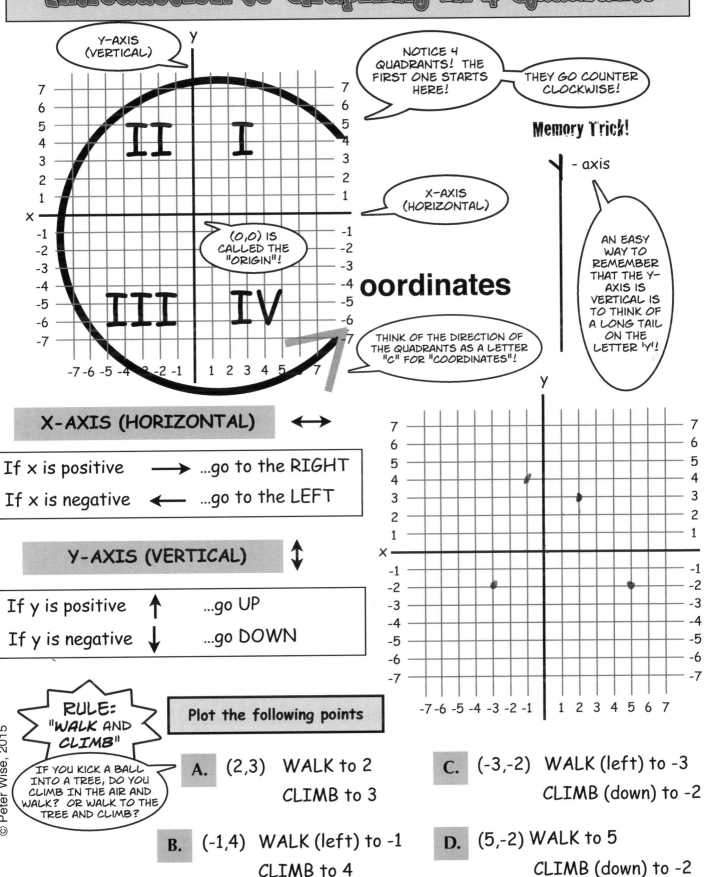

Y-AXIS (VERTICAL)

NOTICE 4 QUADRANTS! THE FIRST ONE STARTS HERE!

THEY GO COUNTER CLOCKWISE!

Memory Trick!

- axis

X-AXIS (HORIZONTAL)

AN EASY WAY TO REMEMBER THAT THE Y-AXIS IS VERTICAL IS TO THINK OF A LONG TAIL ON THE LETTER 'Y'!

(0,0) IS CALLED THE "ORIGIN"!

oordinates

THINK OF THE DIRECTION OF THE QUADRANTS AS A LETTER "C" FOR "COORDINATES"!

X-AXIS (HORIZONTAL) ↔

If x is positive ⟶ ...go to the RIGHT

If x is negative ⟵ ...go to the LEFT

Y-AXIS (VERTICAL) ↕

If y is positive ↑ ...go UP

If y is negative ↓ ...go DOWN

RULE: "WALK AND CLIMB"

IF YOU KICK A BALL INTO A TREE, DO YOU CLIMB IN THE AIR AND WALK? OR WALK TO THE TREE AND CLIMB?

Plot the following points

A. (2,3) WALK to 2
CLIMB to 3

B. (-1,4) WALK (left) to -1
CLIMB to 4

C. (-3,-2) WALK (left) to -3
CLIMB (down) to -2

D. (5,-2) WALK to 5
CLIMB (down) to -2

1

Labeling Points in 4 Quadrants

Plot the following points

1. (3,5) WALK to 3
   CLIMB to 5

2. (-2,4) WALK (left) to -2
   CLIMB to 4

3. (5,-6) WALK to 5
   CLIMB (down) to -6

4. (-4,-3)

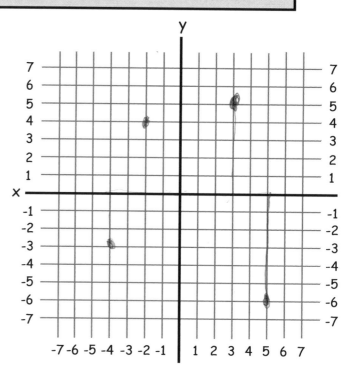

Give the coordinates for the following points

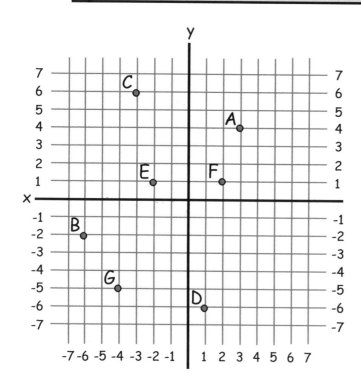

5. Point A (3 , 5)

6. Point B (-6, -2)

7. Point C (-3 , 6)

8. Point D (1, -6)

9. Point E (-2, 1)

10. Point F (2, 1)

11. Point G (-4, 5)

1. Label the points in the four quadrants. What do you notice about the relationship between the x- and y-values in quadrants that are diagonal to each other?

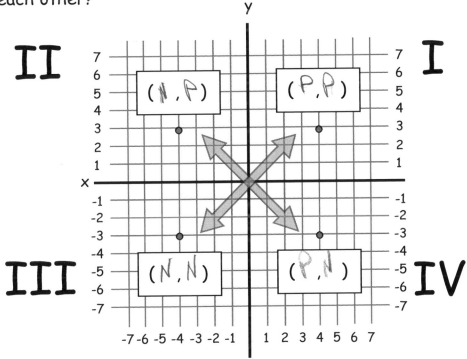

II (N, P)   (P, P) **I**

III (N, N)   (P, N) **IV**

2. Circle the signs that the x- and y-values will have in each quadrant

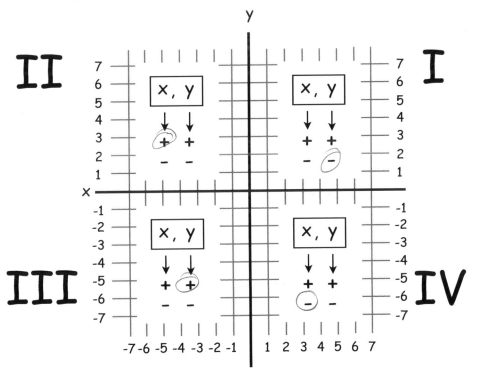

3

What are Intercepts?

A.

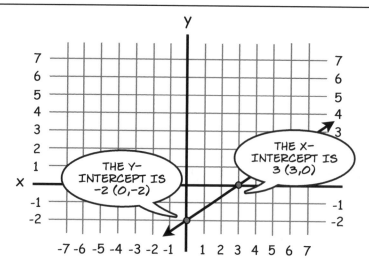

Intercepts are the points where lines cross through the x- or y-axis

Intercepts always have 0 as the other coordinate. This will be important to remember later on.

SUMMARY:

x-intercept = the point on the on the line where the line crosses the x-axis.

(Notice that the y-value for this will always be 0.)

y-intercept = the point on the on the line where the line crosses the y-axis.

(Notice that the x-value for this will always be 0.)

Plot the points and draw the lines

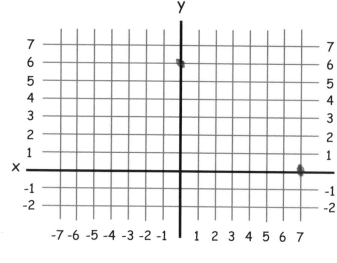

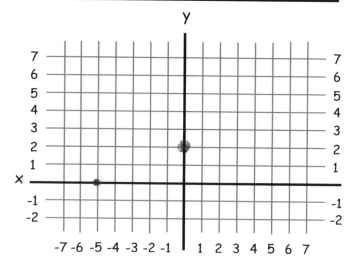

y-intercept = 6

point = (0,6)

x-intercept = 7

point = (7,0)

y-intercept = 2

point (0 , 2)

x-intercept = -5

point (5 , 0)

4

Introduction to Slope

Most common equation for a line:

A. $y = mx + b$

THINK OF THE *M* IN "MOUNTAIN SLOPE"

THE "M" IN THIS EQUATION STANDS FOR THE SLOPE!

IT'S THE SLANT OF THE LINE! NOTICE THAT IT MULTIPLES THE X!

Look at two different points in relation to each other. Measure how you get from one to the other.

When you go from one point of a line to another point:

YOU CAN THINK OF "DOWN" AS UP A NEGATIVE AMOUNT!

$$\text{Slope is } \frac{\text{Rise}}{\text{Run}} = \frac{\text{Up or down how much?}}{\text{Right or left how much?}}$$

RIGHT = POS. LEFT = NEG.

Find the slope of the following lines

$$\frac{\text{Up or down} \updownarrow}{\text{Left or right} \longleftrightarrow}$$

Slope is commonly written as a fraction

1.

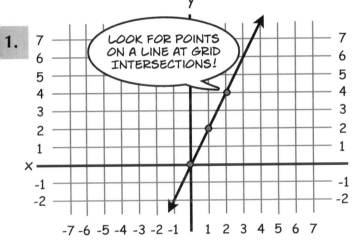

LOOK FOR POINTS ON A LINE AT GRID INTERSECTIONS!

#1 Look for points at GRID INTERSECTIONS

#2 Make a FRACTION:

   a) Numerator: How much the line goes up/down

      Usually it's easiest to measure points going from left to right; but either way is okay!

   b) Denominator: How much the line goes sideways (left/right)

$$\text{Slope} = \frac{\text{Rise}}{\text{Run}} = \frac{\boxed{4} \updownarrow}{\boxed{2} \longleftrightarrow}$$

2.

#1 Look for places where the line crosses at GRID INTERSECTIONS. Plot points here.

#2 $\dfrac{\boxed{5} \updownarrow}{\boxed{4} \longleftrightarrow} \quad \dfrac{\text{Rise}}{\text{Run}}$ **this is the slope**

5

Slope Practice

Find the slope of the following lines

For now, start at the point farthest left

1.

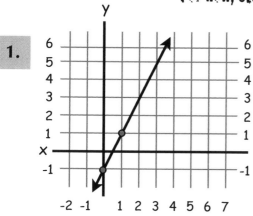

Slope =

Up $\boxed{2}$

Over $\boxed{1}$

4.

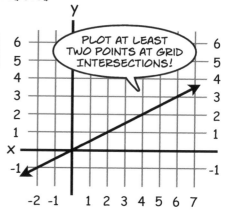

PLOT AT LEAST TWO POINTS AT GRID INTERSECTIONS!

Slope =

$\dfrac{\boxed{1}}{\boxed{2}}$

2.

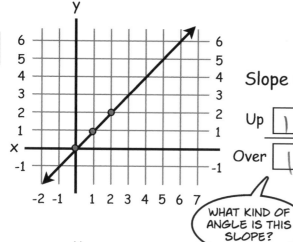

Slope =

Up $\dfrac{\boxed{1}}{\boxed{1}}$ or $\boxed{1}$

Over

WHAT KIND OF ANGLE IS THIS SLOPE?

5.

Slope =

$\dfrac{\boxed{0}}{\boxed{3}}$ or $\boxed{0}$

WHAT IS THE SLOPE OF ANY HORIZONTAL LINE?

3.

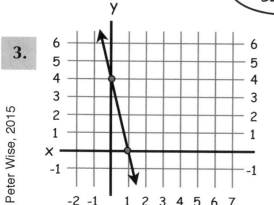

Slope =

$\dfrac{\boxed{-4}}{\boxed{1}}$

WHAT IS DIFFERENT ABOUT THE SLOPE OF THIS LINE?

6.

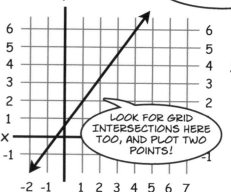

LOOK FOR GRID INTERSECTIONS HERE TOO, AND PLOT TWO POINTS!

Slope =

$\dfrac{\boxed{4}}{\boxed{3}}$

6

Plotting Points and Drawing

Plot points and connect them to draw lines

1.

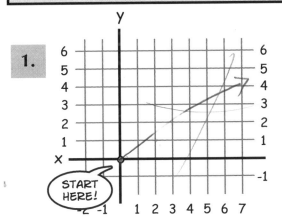

START HERE!

Start at the origin and plot
3 points with a slope of $\frac{1}{2}$

$$y = \frac{1}{2}x$$

slope = $\frac{1}{2}$ $\begin{matrix}\text{(up 1)}\\\text{(right 2)}\end{matrix}$

2.

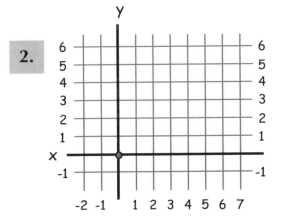

Start at the origin and plot
2 points with a slope of $\frac{2}{3}$

$$y = \frac{2}{3}x$$

slope = $\frac{2}{3}$ $\begin{matrix}\text{(up 2)}\\\text{(right 3)}\end{matrix}$

3.

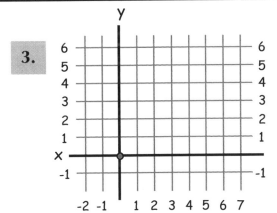

$$y = \frac{3}{1}x$$

Start at the origin and plot
2 points with a slope of $\frac{3}{1}$

4.

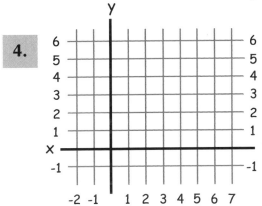

$$y = 5x$$

Start at the origin and plot 1
more point with a slope of 5

FIRST WRITE THE NUMBER 5 AS A FRACTION!

5.

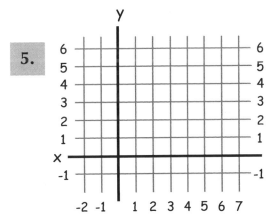

$$y = \frac{2}{7}x$$

Start at the origin and plot
1 more point with a slope of $\frac{2}{7}$

7

Slope-Intercept Form

Plot points and draw lines

1. y-intercept 2   slope $\frac{1}{4}$

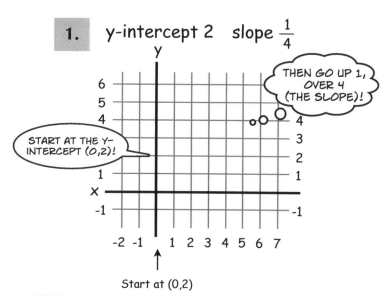

THEN GO UP 1, OVER 4 (THE SLOPE)!

START AT THE Y-INTERCEPT (0,2)!

Start at (0,2)

2. y-intercept -1   slope $\frac{5}{1}$

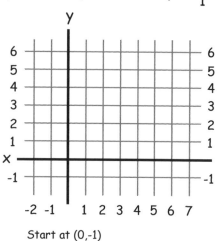

Start at (0,-1)

3. y-intercept 3   slope $\frac{1}{6}$

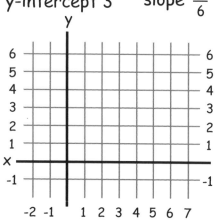

4. y-intercept 1   slope $\frac{4}{5}$

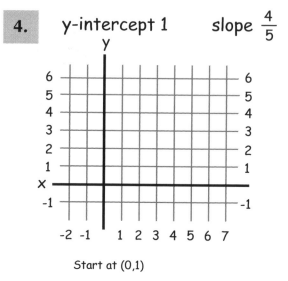

Start at (0,1)

5. y-intercept 0   slope $\frac{5}{3}$

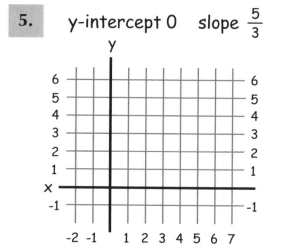

6. y-intercept 3   slope $\frac{2}{7}$

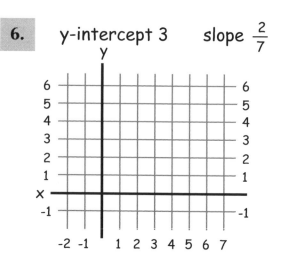

8

Mountain Slope & Balloon

Get acquainted with the various parts of the equation of a line...

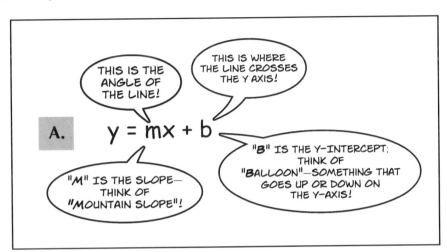

A. $y = mx + b$

THIS IS THE ANGLE OF THE LINE!

THIS IS WHERE THE LINE CROSSES THE Y AXIS!

"M" IS THE SLOPE— THINK OF "MOUNTAIN SLOPE"!

"B" IS THE Y-INTERCEPT; THINK OF "BALLOON"—SOMETHING THAT GOES UP OR DOWN ON THE Y-AXIS!

Line equation:

$y = mx + b$
$y = 1/2x + 2$

↑ slope   ↑ y-intercept

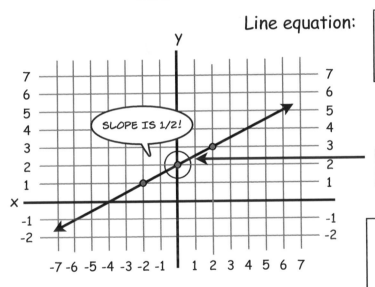

SLOPE IS 1/2!

the y-intercept is 2

REMEMBER! AN INTERCEPT IS WHERE A LINE CROSSES AN AXIS!

Notice that points on the line illustrate equivalent fractions

The slope is $\dfrac{\text{Rise}}{\text{Run}} = \dfrac{1}{2}$ or $\dfrac{2}{4}$ or $\dfrac{3}{6}$

Find the slope and y-intercept of each line:

1. $y = 5x + 3$ slope: 5   y-intercept: 3

2. $y = \dfrac{3}{7}x - 2$ slope: 3/7   y-intercept: -2

3. $y = -\dfrac{4}{9}x - 5$ slope: -4/9   y-intercept: -5

9

Drawing Lines from y = mx + b

How to draw a line with when you have Slope-Intercept form:

1. Put a point on the y-axis for the "b" value.

2. Now look at the "m" value (slope).
   Start at the point you put for "b" and go
   UP and OVER for the "m" value.

3. Draw another point. Connect with a line.

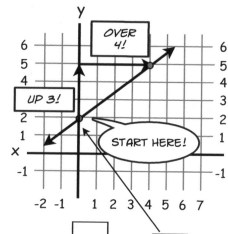

$$y = \begin{matrix}\text{UP}\\\text{OVER}\end{matrix} \boxed{\dfrac{3}{4}} x + \boxed{2}$$

(1) First put a point at (0, 2)

(2) Starting at (0,2) go up 3, over (right) 4 and draw another point.

(3) Connect the points with a line. You're done!

1.

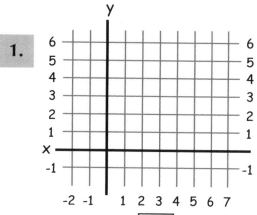

$$y = \begin{matrix}\text{UP}\\\text{OVER}\end{matrix} \boxed{\dfrac{5}{2}} x + \boxed{1}$$

THINK OF "**B**" AS A "**BALLOON**" GOING UP AND DOWN ON THE Y-AXIS!

THINK OF THIS AS YOUR SECRET MAP DIRECTIONS!

3.

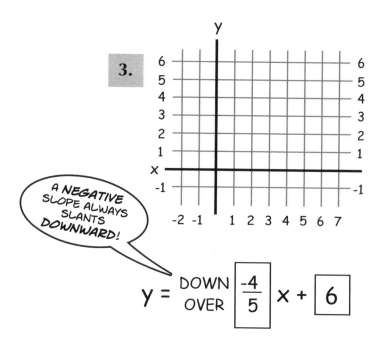

A NEGATIVE SLOPE ALWAYS SLANTS **DOWNWARD!**

$$y = \begin{matrix}\text{DOWN}\\\text{OVER}\end{matrix} \boxed{\dfrac{-4}{5}} x + \boxed{6}$$

2.

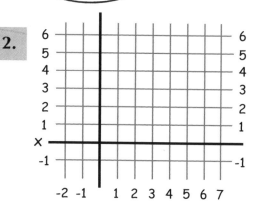

$$y = \begin{matrix}\text{UP}\\\text{OVER}\end{matrix} \boxed{\dfrac{2}{7}} x - \boxed{1}$$

10

Slope Practice, pt. 2

Start at (0, -1) — THIS IS THE "Y-INTERCEPT"!

1.

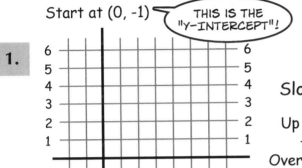

Slope =

Up $\boxed{3}$

Over $\boxed{1}$

Use the slope fraction to plot 2 more points before drawing the line.

4.

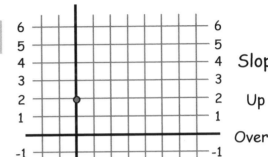

Slope =

Up $\boxed{1}$

Over $\boxed{3}$

Use the slope fraction to plot 2 more points before drawing the line.

TAKE ONE OF THE POINTS YOU ADDED TO THIS GRAPH, AND TRY A SLOPE OF -2/-2.

WHAT DO YOU NOTICE?

2.

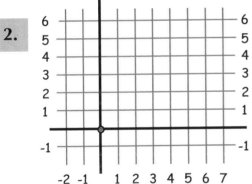

Slope =

Up $\boxed{2}$

Over $\boxed{2}$

Use the slope fraction to plot 2 more points before drawing the line.

WOULD 4/4 GIVE YOU THE SAME SLOPE?

WHY? OR WHY NOT?

5.

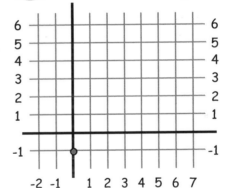

Slope =

Up $\boxed{3}$

Over $\boxed{4}$

Plot only 1 more point before drawing the connecting line.

3.

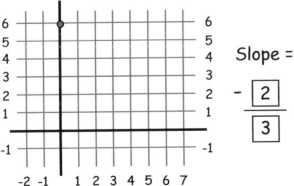

Slope =

$-\dfrac{\boxed{2}}{\boxed{3}}$

Use the slope fraction to plot 2 more points before drawing the line.

6.

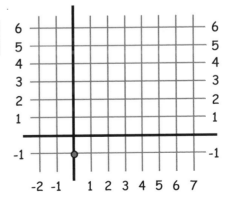

Slope =

$\dfrac{\boxed{6}}{\boxed{5}}$

Plot only 1 more point before drawing the connecting line.

11

Negative Slope

Plot points and draw lines

$$\frac{\text{Negative}}{\text{Negative}} \quad \frac{\text{Rise}}{\text{Run}}$$

Goes DOWN ↓
Goes to the LEFT ←

1.

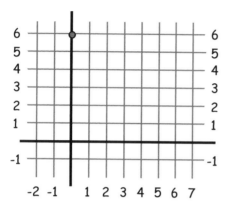

Starting at (0,6), plot two points and connect with a line

$$\text{slope} = \frac{-1 \ (\text{down 1})}{3 \ (\text{right 3})}$$

Now, start with your 2nd point and compare it with this slope:

$$\text{slope} = \frac{1 \ (\text{up 1})}{-3 \ (\text{left 3})}$$

2.

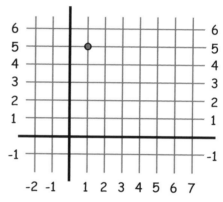

Starting at (1,5), plot two points and connect with a line

$$\text{slope} = \frac{-2}{3}$$

Observation about the negative sign:

3.

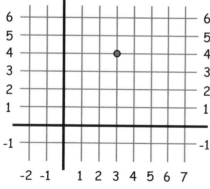

Starting at (3,4), plot two points and connect with a line

$$\text{slope} = \frac{-1}{2}$$

4.

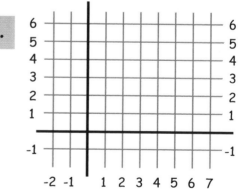

Starting at (-2,5), plot two points and connect with a line

$$\text{slope} = \frac{-3}{2}$$

Does it Matter Where the Negative Sign Goes?

negative sign can go either on the top or bottom

$$\frac{-3}{1} = -3 \qquad\qquad \frac{3}{-1} = -3$$

You can put the negative sign on the numerator

...or the denominator

...same with slope fractions

$$\frac{-2}{3} = \frac{2}{-3}$$

With negative slope fractions you can put the negative sign either on the top or the bottom

YOU'LL GET THE SAME ANSWER EITHER WAY!

Start with the first point; use the slope to draw 1-2 more points

1a.

START HERE. GO DOWN 3 (−3) GO RIGHT 1!

Slope =

Down $\boxed{-3}$

Over $\boxed{1}$

Use the slope to draw one more point

Visual proof that both fractions give the same answer

1b.

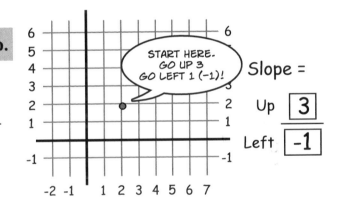

START HERE. GO UP 3 GO LEFT 1 (−1)!

Slope =

Up $\boxed{3}$

Left $\boxed{-1}$

Use the slope to draw one more point

2a.

Slope =

Down $\boxed{-2}$

Over $\boxed{3}$

Start at (-2,6)

Make your first point here; then use the slope to draw two more points

2b.

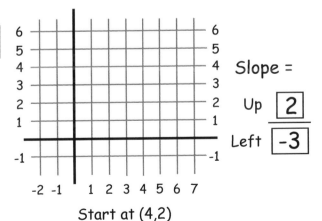

Slope =

Up $\boxed{2}$

Left $\boxed{-3}$

Start at (4,2)

Make your first point here; then use the slope to draw two more points

13

Drawing Rise-Run Arrows

Plot points; draw arrows and lines.

DRAWING SLOPE ARROWS HELPS TO SEE HOW THE LINE IS RELATED TO RISE AND RUN!

A.

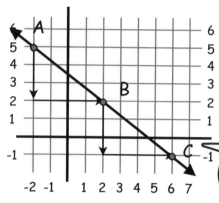

ARROWS HAVE BEEN DRAWN TO SHOW A SLOPE OF (−3)/4!

··· YOUR TURN! DRAW ARROWS SHOWING A SLOPE OF 3/(−4)!

Start at point C and draw arrows to show that you can get the same result by graphing a slope of $\frac{3}{-4}$

slope = $\frac{3}{-4}$ (up 3) (left 4)

1.

Start at (2,3)

Plot one point with a slope of 2/1, draw rise/run arrows (as in Example A), and connect with a line.

slope = $\frac{2}{1}$

Now start at the same point and draw another point having a slope of (-2)/(-1). Draw arrows.

slope = $\frac{-2}{-1}$

2.

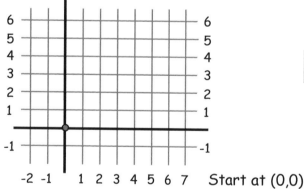

Start at (0,0)

Plot three points, draw rise/run arrows, and connect with a line.

slope = $\frac{2}{2}$

Next, start at (0,0) and draw arrows for these two slopes.

slope = $\frac{3}{3}$ slope = $\frac{5}{5}$

WHAT DO ALL OF THESE FRACTIONS HAVE IN COMMON?

3.

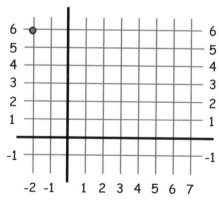

Start at (-2,6)

Plot two points, draw rise/run arrows, and connect with a line.

slope = $\frac{-1}{4}$

14

How Lines are Raised or Lowered

Plot points and connect them to draw lines

Lines on graphs have two parts: (1) Slope (angle), and (2) Height

1.

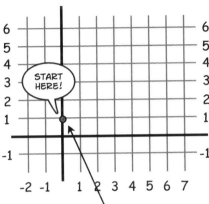

START HERE!

THIS IS CALLED "Y- INTERCEPT"!

$$y = \boxed{\frac{1}{2}} x + 1$$

START UP 1 ON THE Y-AXIS!

#1 Go up 1 on the y-axis and plot your first point (0,1)

#2 Now go up the slope and plot the next point

3.

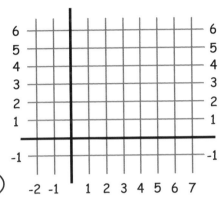

$$y = \frac{3}{4} x + 3$$

#1 Go up this amount on the y-axis and plot your first point

#2 Now go up the slope and plot the next point

2.

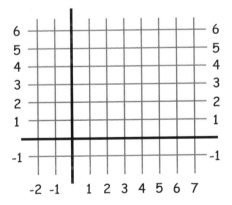

$$y = \boxed{\frac{3}{1}} x + 2$$

HEIGHT OF THE LINE!

#1 Go up 2 on the y-axis and plot your first point

#2 Now go up the slope and plot the next point

ANGLE (SLOPE) OF THE LINE!

4.

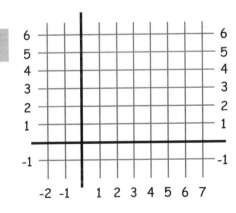

$$y = \frac{4}{5} x - 1$$

THIS IS THE "Y-NTERCEPT"!

#1 Look at the value of the y-intercept and go down this amount

#2 Now go up the slope and plot the next point

15

Plug in for 'm' and 'b'

The way of describing a line is called Slope-Intercept Form

$$y = mx + b$$

$$y = (slope)(x) + (y\text{-intercept})$$

"M" IS THE SLOPE

"B" IS THE Y-INTERCEPT!

A.

THINK OF THE "M" AS "MOUNTAIN SLOPE"

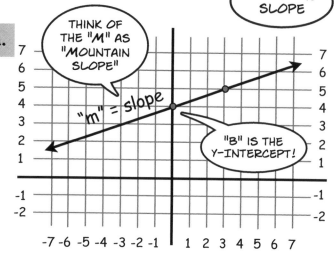

"m" = slope

"B" IS THE Y-INTERCEPT!

Y = LINE SLANT + HOW HIGH ON THE Y-AXIS!

$$y = mx + b$$

$$y = \boxed{\dfrac{1}{3}}\, x + \boxed{4}$$

1.

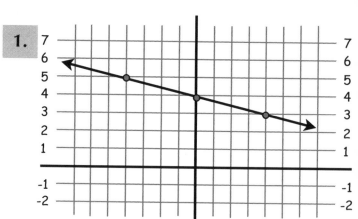

$$y = \boxed{\dfrac{1}{4}}\, x + \boxed{4}$$

Find the slope and y-intercept of the following lines and substitute these values for "m" and "b"

2.

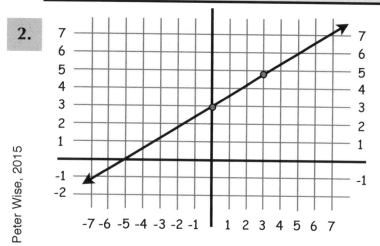

$$y = \boxed{\dfrac{2}{3}}\, x + \boxed{3}$$

3.

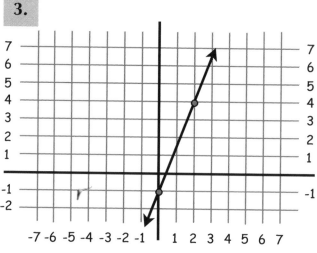

$$y = mx + b \qquad m = \boxed{\dfrac{5}{2}} \quad b = \boxed{-1}$$

16

The Parentheses Trick

- When working with linear equations you will often substitute numbers for variables to plot points
- To substitute a value for a variable—rewrite the equation, PUTTING PARENTHESES in place of that variable (x or y)

PUT PARENTHESES IN PLACE OF THE VARIABLES!

$$y = 3x + 1$$

$$y = 3(\ ) + 1$$

Now pick ANY values for x and solve for y. We started with x = 1

$$y = 3(1) + 1$$

$$y = 4$$

x	y
1	

Q. Why do you want to substitute numbers for x?

A. Because this will give you ordered pairs that are on the line

Parentheses provide a helpful "container" for the substituted numbers

---

Tips for Substituting Numbers

- Plug in any numbers you want for the x-value, and then calculate the y-value
- Often (especially with lines) it's best to plug in x-values that are close to zero: -2, -1, 0, 1, 2, etc.

Rewrite the following equations using parentheses; then plug in two x-values and solve for y

1. $y = 2x + 4$

$$y = 2(\ ) + 4$$

This is how to write the equation using parentheses for the x-value

x	y
0	4
1	6

3. $y = -3x + 6$

x	y
0	6
1	3

2. $y = 5x - 5$

Rewrite the equation using parentheses for the x-value

x	y
0	-5
1	0

4. $y = -1x - 2$

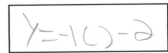

x	y
0	-2
-1	-1

17

Plot by Making an x/y Table

How to draw a line using an x, y table:

1. Make a blank x,y table

2. Pick easy numbers (like 0, 1, 2) for x. Substitute them for x to figure out the corresponding y-values.

3. Plot the two x, y points and connect them with a line

Example

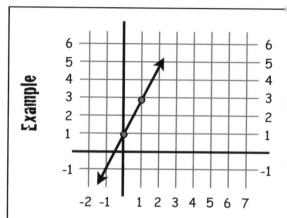

$$y = \boxed{2} x + 1$$

Same as $\boxed{\dfrac{2}{1}}$

x	y
0	1
1	3

1.

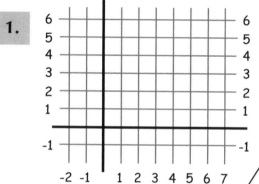

$$y = 3x + 2$$

$y = \ 3(0) \ + \ 2$

$y = \ 3(1) \ + \ 2$

x	y
0	
1	

ACTUALLY, YOU CAN PLUG IN ANY NUMBER FOR X TO FIND ITS CORRESPONDING Y-VALUE!

---BUT SMALLER NUMBERS ARE USUALLY EASIER TO WORK WITH!

FIND THE Y VALUE BY PLUGGING IN THE X VALUE!

3.

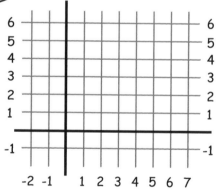

$$y = \frac{1}{5} x + 5$$

x	y
0	
5	

USE A VALUE FOR X THAT WILL CANCEL OUT THE DENOMINATOR!

2.

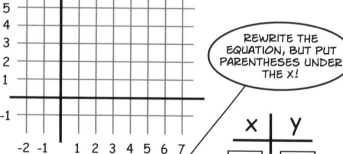

$$y = -2x + 4$$

REWRITE THE EQUATION, BUT PUT PARENTHESES UNDER THE X!

x	y

18

Practice with x/y Tables

Plug in the x-values and solve for y

1. $y = x + 2$

Plug in the x values and see what the y values become!

$y = (\quad) + 2$

↖ Solve for y   Plug in the x values here ↗

USING PARENTHESES AS A CONTAINER FOR SUBSTITUTING NUMBERS IS A GOOD PRACTICE, EXPECIALLY IN THE EARLY STAGES OF DOING SUBSTITUTION!

x	y
-2	0
-1	1
1	3
2	4
8	10

2. same as y = (-1)x

$y = -x$

$y = -(\quad)$

NOTE! NEGATIVE ONE IS A SIGN SWITCHER!

x	y
-2	2
-1	1
1	-1
2	-2

3. $y = \frac{1}{2}x + 1$

$y = \frac{1}{2}(\quad) + 1$

With slopes in fraction form, it is easiest to plug in values that are multiples of the denominator—you will always get whole numbers! (This is because the multiples will cancel the denominator.)

x	y
-6	-2
-2	0
0	1
2	
8	5

4. $y = \frac{1}{3}x - 2$

$y = \frac{1}{3}(\quad) - 2$

x	y
-6	-4
-3	-3
3	-1
9	1
12	2

5. $y = -\frac{3}{5}x + 1$

$y = -\frac{3}{5}(\quad) + 1$

x	y
-10	7
-5	4
10	-5
20	-11
30	-17

6. $y = -x + 1$

$y = -(\quad) + 1$

x	y
-2	3
-1	2
1	0
2	-1
6	-5

19

Practice with x/y Tables

Plug in the x-values and solve for y

1. $y = 2x + 1$

$(\ ) = 2(\ ) + 1$

x	y
-2	
-1	
0	
1	
2	

4. $y = 5x - 10$

x	y
-4	
-2	
0	
2	
4	

2. $y = -5x$

x	y
-2	
-1	
1	
2	

USE THE PARENTHESES TRICK!

REMEMBER! MULTIPLYING BY A NEGATIVE NUMBER SWITCHES ALL OF THE SIGNS!

5. $y = \frac{2}{3}x + 3$

x	y
-6	
-3	
0	
3	
6	

With slopes in fraction form, it is easiest to plug in values that are multiples of the denominator—you will always get whole numbers! (This is because the multiples will cancel the denominator.)

3. $y = 3x - 2$

x	y
-2	
-1	
0	
1	
2	

6. $y = \frac{4}{5}x - 2$

x	y
-10	
-5	
0	
5	
10	

20

Substitution with Fraction Slopes

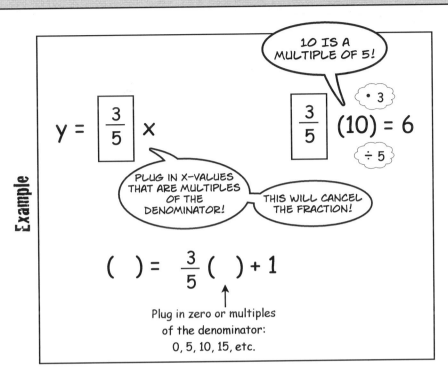

Example

$y = \boxed{\dfrac{3}{5}} x$

PLUG IN X-VALUES THAT ARE MULTIPLES OF THE DENOMINATOR!

THIS WILL CANCEL THE FRACTION!

10 IS A MULTIPLE OF 5!

$\boxed{\dfrac{3}{5}} (10) = 6$

· 3
÷ 5

$(\quad) = \dfrac{3}{5} (\quad) + 1$

Plug in zero or multiples of the denominator:
0, 5, 10, 15, etc.

Rewrite the following equations using parentheses; then plug in two x-values that are MULTIPLES of the DENOMINATOR, and solve for y

1. $y = \dfrac{2}{3}x + 1$

x	y

Rewrite the equation using parentheses for the x-values; plug in multiples of the denominator (3, 6, 9, 12, etc)

3. $y = \dfrac{5}{7}x - 2$

x	y

2. $y = \dfrac{3}{4}x + 2$

x	y

4. $y = -\dfrac{2}{5}x + 3$

x	y

The y-in-Terms-of-x Trick

Copy the expression that the y term equals and use it instead of the y-value in your table

Just copy the entire side opposite the y (only if the y is by itself)

Example

A. $y = 7x + 2$

↑   ↑
Both sides are equal

x	(y) (= 7x + 2) 7( ) + 2
0	☐
1	☐

THIS IS THE SAME AS Y IN THIS EQUATION, SO YOU CAN JUST WRITE IT HERE AND SOLVE NORMALLY!

Make x/y tables, but now write y in terms of x (as in the example above)

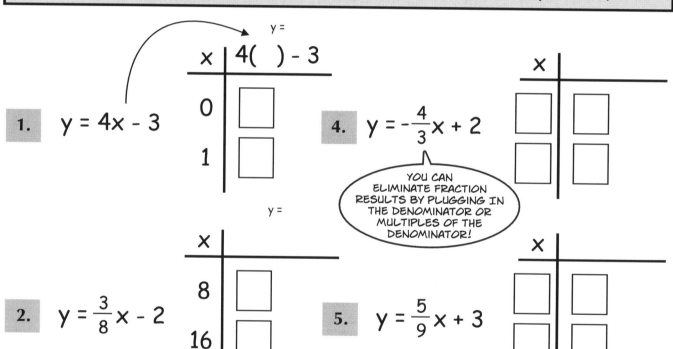

1. $y = 4x - 3$

x	y = 4( ) - 3
0	☐
1	☐

4. $y = -\frac{4}{3}x + 2$

YOU CAN ELIMINATE FRACTION RESULTS BY PLUGGING IN THE DENOMINATOR OR MULTIPLES OF THE DENOMINATOR!

2. $y = \frac{3}{8}x - 2$

x	y =
8	☐
16	☐

5. $y = \frac{5}{9}x + 3$

3. $y = -\frac{7}{2}x - 1$

x	
2	☐
4	☐

6. $y = -\frac{8}{5}x + 10$

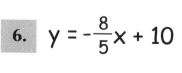

Seeing y-in-Terms-of-x on a Graph

Writing y as a function of x; A.K.A "f of x"

1. $y = 2x$

IN THIS EQUATION, THE Y-VALUES ARE WHAT YOU GET WHEN YOU MULTIPLY X BY 2!

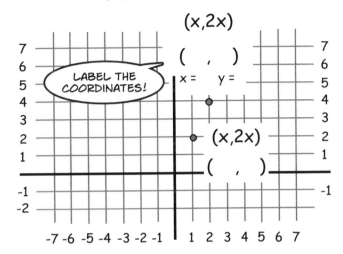

(x,2x)

(,)

x = y =

LABEL THE COORDINATES!

(x,2x)

(,)

2. $y = \frac{1}{2}x$

THE HEIGHT (Y-VALUE) WILL BE 1/2 OF THE SIDEWAYS DIRECTION (X-VALUE)!

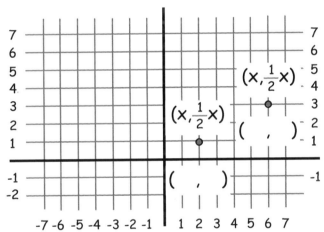

$(x,\frac{1}{2}x)$

$(x,\frac{1}{2}x)$

(,)

(,)

3. $y = 3x - 2$

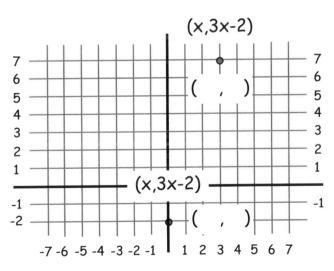

(x,3x-2)

(,)

(x,3x-2)

(,)

WRITE THE COORDINATES AS ABOVE (Y IN TERMS OF X)

4. $y = 5x$ (,)

5. $y = \frac{4}{5}x + 3$ (,)

6. $y = -\frac{1}{4}x - 2$ (,)

23

© Peter Wise, 2015

Graph and Explain the Difference

What differences do you see between these pairs of linear equations?

1a.   y = 1x + 0

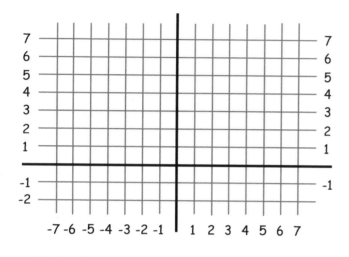

1b.   y = 1x + 2

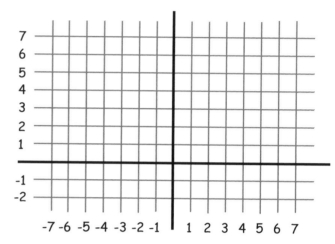

Explain the difference:

2a.   y = 2x + 2

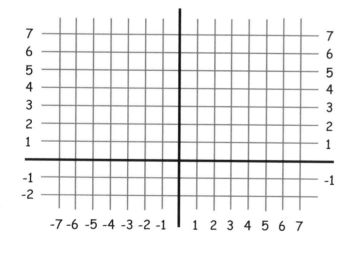

2b.   y = 2x - 2

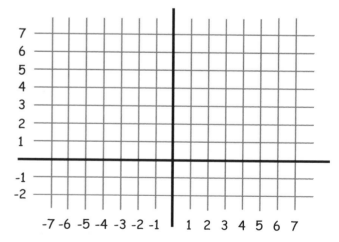

Explain the difference:

Graph and Explain the Difference

Graph the lines and make an observation about the differences in slope

1.   y = 6x + 0

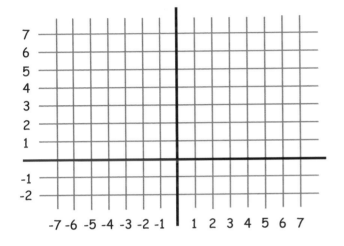

3. $y = \frac{1}{2}x + 0$

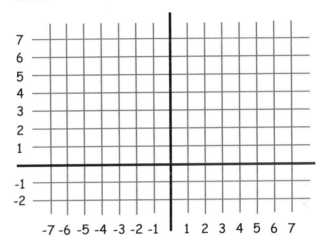

2.   y = 1x + 0

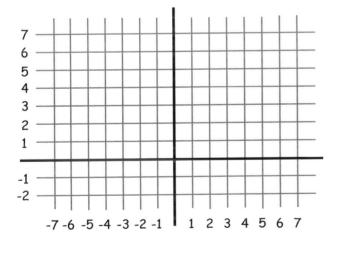

4. $y = \frac{1}{7}x + 0$

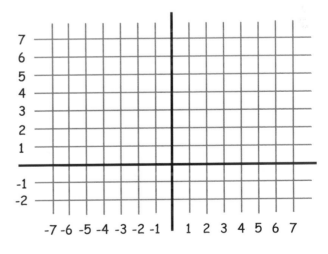

What do you notice in the lines as you go from problem 1 to problem 4?

Graph and Explain the Difference

Plot the points and draw the lines

1a.   y = 3x + 1

1b.   y = -3x + 1

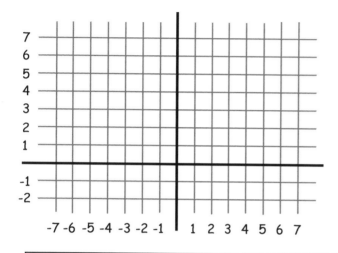
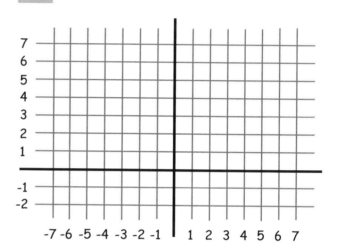

Explain the difference:

2a.   y = 2x + 1   **2b.** $y = -\frac{1}{2}x + 1$

Plot points and draw lines on the same graph

3a. $y = \frac{2}{3}x + 1$ **3b.** $y = -\frac{3}{2}x + 1$

Plot points and draw lines on the same graph

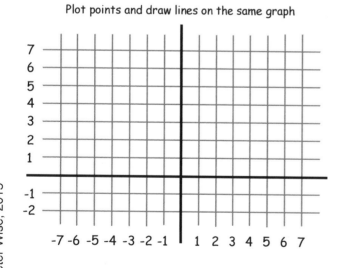
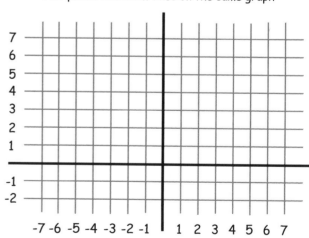

Explain the difference? Be specific!

Slope and y-Intercept

Find the slope and the y-intercept of each equation

1. $y = \dfrac{2}{3}x + 5$ slope = $\boxed{\dfrac{2}{3}}$ y-intercept = $\boxed{5}$

2. $y = \dfrac{5}{4}x - 2$ slope = $\boxed{}$ y-intercept = $\boxed{}$

3. $y = \dfrac{6}{1}x - 7$ slope = $\boxed{}$ y-intercept = $\boxed{}$

4. $y = \dfrac{8}{9}x - 1$ slope = $\boxed{}$ y-intercept = $\boxed{}$

5. $y = \boxed{} \, x \, \boxed{+}$ slope = $\boxed{\dfrac{8}{3}}$ y-intercept = $\boxed{3}$

6. $y = \dfrac{-4}{3}x + 8$ slope = $\boxed{}$ y-intercept = $\boxed{}$

7. $y = \boxed{} \, x \, \boxed{}\boxed{}$
$\uparrow$
put + or –
here
    slope = $\boxed{\dfrac{7}{-5}}$ y-intercept = $\boxed{-1}$

8. $y = \dfrac{7}{-11}x$ slope = $\boxed{}$ y-intercept = $\boxed{}$

9. $y = \boxed{} \, x \, \boxed{}\boxed{}$ y-intercept = 4   slope = $\dfrac{1}{5}$

10. $y = \boxed{} \, x \, \boxed{}\boxed{}$ b = -15   m = -6

Slope & y-Intercept from Graphs

Determine the slope and the y-intercept from the graphs

SLOPE
goes here

y-intercept
goes here

1.
$$y = \boxed{} \times \boxed{} \boxed{}$$

put + or - here

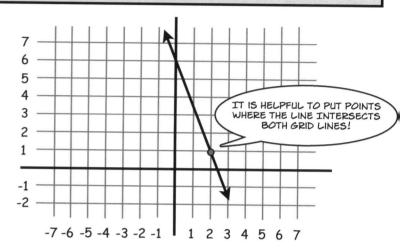

IT IS HELPFUL TO PUT POINTS WHERE THE LINE INTERSECTS BOTH GRID LINES!

2.
$$y = \boxed{} \times \boxed{} \boxed{}$$

put + or - here

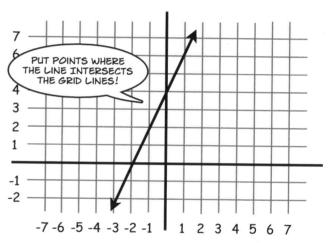

PUT POINTS WHERE THE LINE INTERSECTS THE GRID LINES!

3.
$$y = \boxed{} \times \boxed{} \boxed{}$$

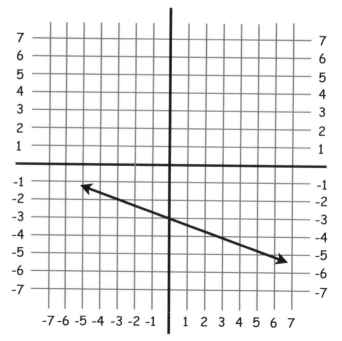

To make a line perpendicular to this one, just change the slope:

$$y = \frac{3}{4}x + 2$$

↑

• Use the RECIPROCAL of this slope
• Make it NEGATIVE

↓

$$y = -\frac{4}{3}x + 2$$

THE FRACTION IS FLIPPED AND IT HAS THE OPPOSITE SIGN!

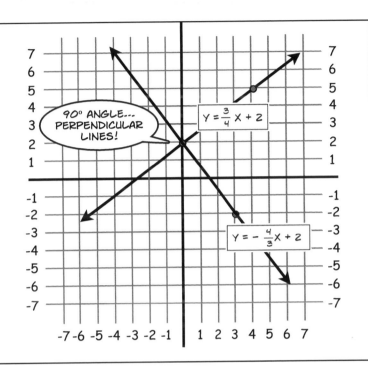

90° ANGLE... PERPENDICULAR LINES!

$Y = \frac{3}{4}x + 2$

$Y = -\frac{4}{3}x + 2$

Give an equation for a perpendicular line

1. $y = \frac{2}{5}x + 1$

4. $y = -2x + 5$

2. $y = 3x + 4$

5. $y = -1\frac{1}{3}x - 2$

3. $y = -\frac{7}{3}x + 2$

6. $y = 2\frac{3}{4}x + 1$

If the Coefficient of y Isn't 1

Make the y-value 1y

Example

A. $2y = 6x + 8$

THE Y-VALUE HAS TO BE ONE—SO GET RID OF THE 2 BY DIVIDING EVERY TERM BY 2!

$\dfrac{2y}{2} = \dfrac{6x}{2} + \dfrac{8}{2}$ ⟶ $\boxed{y = 3x + 4}$

Rewrite the equations so that the y-value is 1y (or just y)

THIS HAS TO BE 1Y!

1. $5y = 10x + 15$

Divide every term by: ☐

Rewritten equation

2. $3y = x + 12$

Divide every term by: ☐

Rewritten equation

3. $-y = 2x + 10$

THERE IS AN INVISIBLE NUMBER IN FRONT OF THE Y!

Divide every term by: ☐
(or multiply)

Rewritten equation

4. $\dfrac{3}{4}y = 15x + 9$

WHAT MAGIC TOOL MAKES FRACTIONS DISAPPEAR BY TURNING THEM TO ONES?

Multiply every term by: ☐

Rewritten equation

y = mx + b is Slope-Intercept Form—you can arrange the equation to match this form

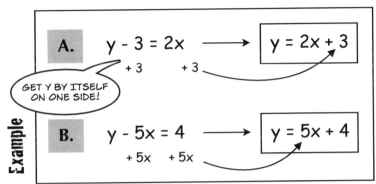

Example

A. y - 3 = 2x → y = 2x + 3
   + 3   + 3

GET Y BY ITSELF ON ONE SIDE!

B. y - 5x = 4 → y = 5x + 4
   + 5x   + 5x

Change sides, change signs

Negative 3 on one side of the equal sign = positive 3 on the other side

- 5x on one side of the equal sign = + 5x on the other side

Rewrite the equations so that they are in slope-intercept form (y = mx + b)

1. y - 10 = 4x

2. y - 3 = 4x + 5

3. $y - \frac{2}{5} = 7x + \frac{4}{5}$

4. $y + 6 = \frac{3}{7}x + 4$

5. y + 5x = -2x - 6

6. y + 7x = 5x + 1

7. y + 10x - 8 = 5x - 10

8. y - x - 4 = 0

9. -3x + 7 + y = -11x + 2

10. $y + \frac{3}{7}x = \frac{2}{3}x - 4$

A. Find the slope of the line going through these points:

$(8,5)$, $(3,2)$

Example

#1 Circle the y-value (second number) in each ordered pair

$(8,\textcircled{5}), (3,\textcircled{2})$

(THE Y-VALUES ARE ON THE RIGHT SIDE!)

TO SEE THE SLOPE CLEARLY, YOU NEED THE RIGHT GLASSES!

#2 Put these numbers in the ovals (at the TOP of the fraction) and subtract them

The RIGHT numbers wear the GLASSES on the TOP ("head") of the fraction!

$$\frac{\boxed{5} - \boxed{2}}{\boxed{8} - \boxed{3}} = \frac{3}{5}$$

#3 Put the x-values on the BOTTOM and subtract them

Find the slope of the following points by SUBTRACTION; use the GLASSES TRICK

THE "GLASSES" ARE CONTAINERS THAT HELP YOU KEEP TRACK OF NEGATIVE NUMBERS AND SUBTRACTION SIGNS!

1. Find the slope of the line going through these points: $(4,6)$, $(1,5)$

HAVING THE SIGNS OF THE NEGATIVE INTEGERS IN HERE HELPS TO KEEP TRACK OF THE SIGNS!

• Circle the y-values
• Put them in the "glasses frames"
• Put the x-values on the bottom boxes

• Subtract top and bottom to find the slope (often it will be a fraction)

ACTUALLY, YOU CAN SUBTRACT IN THE REVERSE DIRECTION AND STILL GET THE SAME ANSWER!

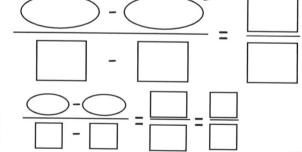

2. Find the slope of the line going through these points: $(3,2)$, $(1,1)$

• Circle the y-values (RIGHT numbers)
• Put these "glasses numbers" on top

DRAW GLASSES FOR THE TOP NUMBERS AND BOXES FOR THE BOTTOM NUMBERS!

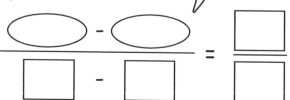

Finding Slope from Two Points

Use the GLASSES TRICK to find the slope of the following points

1. Find the slope of the line going through these points: $(4,2),(1,1)$

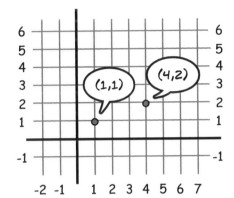

#1 Circle the y-value (second number) in each ordered pair

#2 Put these numbers in the ovals (numerators) and subtract them

#3 Put the x-values on the bottom and subtract them

It doesn't matter which point you start with as long as you are consistent

$(4,\textcircled{2}),(1,\textcircled{1})$

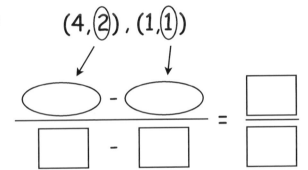

2. Find the slope of the line going through these points: $(6,9),(2,6)$

CONTINUE USING THE **GLASSES TRICK!**

3. Find the slope of the line going through these points: $(3,8),(6,3)$

4. Find the slope of the line going through these points: $(-7,-6),(-4,-2)$

THE GLASSES FRAMES HELP YOU KEEP TRACK OF NEGATIVE NUMBERS AND SUBTRACTION SIGNS!

33

Finding Slope from Two Points

Use the GLASSES TRICK to find the slope from two points

---

1. Find the slope of the line going through these points: $(3,5), (2,1)$

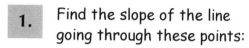

Circle the y-values

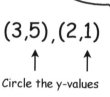

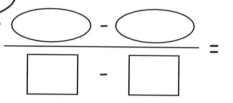

Y-VALUES GO ON TOP!

The x-values go on the bottom

2. Find the slope of the line going through these points: $(4,1), (11,4)$

You can subtract the first point minus second the point or the second point minus the first point, as long as you are consistent

REMEMBER THAT "MINUS A NEGATIVE" = "ADD A POSITIVE"!

3. Find the slope of the line going through these points: $(-1,-14), (-2,10)$

DRAWING THESE FRAMES HELPS PREVENT LOSING TRACK OF NEGATIVE SIGNS!

4. Find the slope of the line going through these points: $(-7,-4), (-2,7)$

5. Find the slope of the line going through these points: $(-3,2), (12,-5)$

34

Finding Slope from Two Points

	Find the slope of the following points by SUBTRACTION

1. Find the slope of the line going through these points: $(4,1)$, $(7,5)$

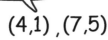

CIRCLE THE Y-VALUES!

$$\frac{\bigcirc - \bigcirc}{\square - \square} = \frac{\square}{\square}$$

2. Find the slope of the line going through these points: $(-5,6),(-3,3)$

DRAW IN THE OVAL AND RECTANGLE FRAMES!

$$\frac{-}{-} = \frac{\square}{\square}$$

3. Find the slope of the line going through these points: $(-6,-3),(2,4)$

$$\frac{}{} = \frac{\square}{\square}$$

4. Find the slope of the line going through these points: $(5,-4),(-5,1)$

$$\frac{}{} = \frac{\square}{\square}$$

5. Find the slope of the line going through these points: $(-9,-7),(-4,-1)$

$$\frac{}{} = \frac{\square}{\square}$$

6. Find the slope of the line going through these points: $(-7,-2),(4,-5)$

$$\frac{}{} = \frac{\square}{\square}$$

Finding Slope from Tables

1. Find the slope of the line going through these points:

REMEMBER! ONLY Y-VALUES GO HERE ON TOP!

$$\frac{\bigcirc - \bigcirc}{\square - \square} = \frac{\square}{\square}$$

THE TEMPTATION IS TO PUT THE NUMBERS IN THE TOP ROW ON TOP OF THE SLOPE FRACTION, INSIDE THE GLASSES—BUT DON'T DO THIS IF THE TOP ROW HAS THE X-VALUES!

x	-4	-2	0	2
y	10	3	-4	-11

subtract any of these on the BOTTOM

y-values can appear on either the top or bottom rows; be careful!

subtract any of these on the TOP

FIRST CHECK TO SEE WHERE THE Y-VALUES ARE! PUT THEM ON TOP, NO MATTER WHERE THEY APPEAR IN A TABLE!

2. Find the slope of the line going through these points:

y	14	4	-1	-6
x	-8	4	10	16

$$\frac{\bigcirc - \bigcirc}{\square - \square} = \frac{\square}{\square} = -\frac{\square}{\square}$$

simplify

3. Find the slope of the line going through these points:

x	-9	1	11	21
y	-10	-7	-4	-1

$$\frac{\bigcirc - \bigcirc}{\square - \square} = \frac{\square}{\square}$$

4. Find the slope of the line going through these points:

y	-22	-11	0	11
x	-16	-10	-4	2

$$\frac{\bigcirc - \bigcirc}{\square - \square} = \frac{\square}{\square} = \frac{\square}{\square}$$

36

Linear Equations Review

1. Find the slope of the line going through these points: (-6,3),(-4,-2)

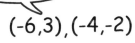

CIRCLE THE Y–VALUES!

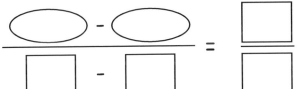

DRAW IN THE OVAL AND RECTANGLE FRAMES!

2. Find the slope of the line going through these points: (-9,-7),(-6,-5)

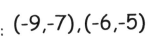 =

3. Plot three points (one x-value can be zero)

$$y = \frac{1}{5}x + 1$$

$$(\quad) = \frac{1}{5}(\quad) + 1$$

CHOOSE X-VALUES THAT ARE MULTIPLES OF 5 TO ELIMINATE FRACTIONS!

x	y

4. Negative slopes always go:

Upward   Downward

(Circle one)

...as you move from left to right on a graph

5. Graph the line: y = -2/3x + 4

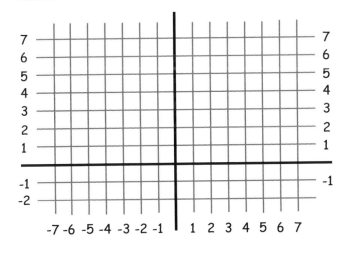

6. Give the equation of the line

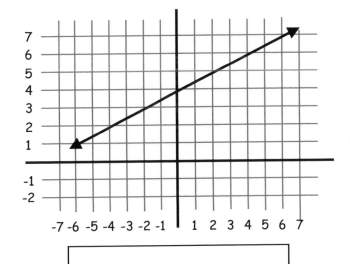

© Peter Wise, 2015

37

A. Find the equation of the line with a slope of 2 and going through the point (3,10)

Example

PICK ANY POINT, USE ITS X- AND Y-VALUES!

x = 3

THE SLOPE IS GIVEN TO YOU IN THE PROBLEM!

y = 10

m = 2

USE ALL OF THE ABOVE NUMBERS TO SOLVE FOR THIS!

b = ☐

$$y = mx + b$$
$$(10) = (2)(3) + b$$

(3,10) (the point we are given)

$$10 = 6 + b$$
$$b = 4$$

Equation: y = 2 x + 4

Y = MX + B HAS FOUR PIECES OF INFORMATION. IF YOU HAVE THREE VALUES, YOU CAN FIGURE OUT THE FOURTH VALUE!

Find the equations of the following lines

1. Find the equation of the line with a slope of 1/2 and going through the point (4,5)

x = ☐   m = ☐

PLUG IN THE VALUES!

$$y = mx + b$$
$$( ) = ( )( ) + b$$

LINEAR EQUATIONS NEED TO HAVE JUST "X" AND "Y" AS VARIABLES!

y = ☐   b = ☐

IF YOU HAVE THREE VALUES YOU CAN FIGURE OUT THE FOURTH!

Equation: y = ☐ x + ☐

2. Find the equation of the line with a slope of 3 and going through the point (1,7)

x = ☐   m = ☐

$$( ) = ( )( ) + b$$

y = ☐   b = ☐

Equation: y = ☐ x + ☐

3. Find the equation of the line with a slope of 2 and going through the point (-5,8)

x = ☐   m = ☐

$$( ) = ( )( ) + b$$

y = ☐   b = ☐

Equation: y = ☐ x + ☐

Parallel Lines Have the Same Slope

© Peter Wise, 2015

Example

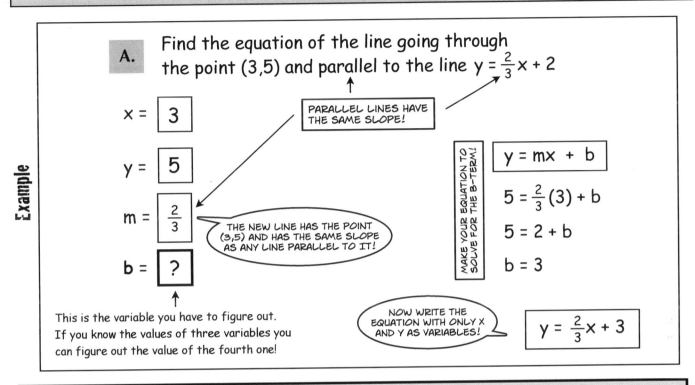

A. Find the equation of the line going through the point (3,5) and parallel to the line $y = \frac{2}{3}x + 2$

PARALLEL LINES HAVE THE SAME SLOPE!

$x = \boxed{3}$

$y = \boxed{5}$

$m = \boxed{\frac{2}{3}}$

THE NEW LINE HAS THE POINT (3,5) AND HAS THE SAME SLOPE AS ANY LINE PARALLEL TO IT!

$b = \boxed{?}$

This is the variable you have to figure out. If you know the values of three variables you can figure out the value of the fourth one!

MAKE YOUR EQUATION TO SOLVE FOR THE B-TERM!

$y = mx + b$

$5 = \frac{2}{3}(3) + b$

$5 = 2 + b$

$b = 3$

NOW WRITE THE EQUATION WITH ONLY X AND Y AS VARIABLES!

$y = \frac{2}{3}x + 3$

Write equations for the following lines (solve for a new "b" value)

1. Find the equation of the line going through the point (12,7) and parallel to the line $y = \frac{1}{3}x + 8$

$x = \boxed{}$ $m = \boxed{}$

$y = \boxed{}$ $b = \boxed{}$

$y = \boxed{}x + \boxed{}$

2. Find the equation of the line going through the point (-8,10) and parallel to the line $y = -\frac{3}{4}x + 4$

$x = \boxed{}$ $m = \boxed{}$

$y = \boxed{}$ $b = \boxed{}$

$\boxed{}$

39

Parallel Lines Have the Same Slope

1. Find the equation of the line going through the point (-6,-1) and parallel to the line $y = -\frac{2}{3}x - 3$

x = ☐   m = ☐

y = ☐   b = ☐ ← This is the variable you have to figure out, given the slope and the points (-6,-1)

Equation:

☐

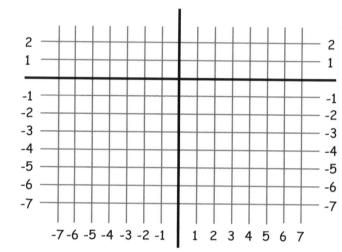

GRAPH BOTH PARALLEL LINES TO SEE VISUALLY WHAT IS GOING ON!

2. Find the equation of the line going through the point (-10,10) and parallel to the line $y = -\frac{3}{5}x - 5$

x = ☐   m = ☐

y = ☐   b = ☐

Equation:

☐

3. Find the equation of the line going through the point (-8,-7) and parallel to the line $y = -\frac{3}{2}x - 2$

Make a list of the four variables as above

↓

NOTE! ANSWERS AREN'T ALWAYS WHOLE NUMBERS!

Equation:

☐

Finding Equations of Perpendicular Lines

© Peter Wise, 2015

Example

A. Find the equation of the line going through the point (-2,4) and perpendicular to the line $y = \frac{2}{3}x + 1$

THE SLOPES OF PERPENDICULAR LINES ARE NEGATIVE RECIPROCALS

$x = \boxed{-2}$

$y = \boxed{4}$

$m = \boxed{-\frac{3}{2}}$

(a) Flip the fraction (= make it a reciprocal)

(b) Switch the sign

$b = \boxed{?}$

↑ This is the variable you have to figure out; if you have three variables you can figure out the fourth one!

MAKE YOUR EQUATION TO SOLVE FOR THE B-TERM!

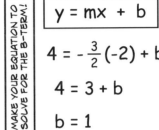

$y = mx + b$

$4 = -\frac{3}{2}(-2) + b$

$4 = 3 + b$

$b = 1$

NOW WRITE THE EQUATION WITH ONLY X AND Y AS VARIABLES!

$y = -\frac{3}{2}x + 1$

Write equations for the following lines

1. Find the equation of the line going through the point (5,6) and perpendicular to the line $y = -\frac{5}{4}x + 1$

$x = \boxed{}$ $m = \boxed{}$

(a) Flip the slope fraction (= make it a reciprocal)

(b) Switch the sign

$y = \boxed{}$ $b = \boxed{}$

Equation:

$\boxed{}$

2. Find the equation of the line going through the point (-3,12) and perpendicular to the line $y = \frac{3}{7}x - 6$

REMEMBER TO GIVE THE SLOPE OF THE NEW (PERPENDICULAR) LINE!

$x = \boxed{}$ $m = \boxed{}$

$y = \boxed{}$ $b = \boxed{}$

Equation:

$\boxed{}$

41

Variables and Constants in y = mx + b

These are constants (fixed numbers that don't change)

$$y = \boxed{m}x + \boxed{b}$$

YOU CAN PLUG IN AN X- OR Y-VALUE FROM ANY POINT ON THE LINE TO HELP YOU FIGURE OUT THE "M" OR "B" VALUES!

x and y stay variables; they represent infinite pairs of solutions (all on the line)

THE X- AND Y-VALUES ON THE LINE CAN EVEN BE FRACTIONS BETWEEN THE WHOLE NUMBERS ON THE LINE—THEY WILL STILL MAKE THE EQUATION TRUE!

These numbers are fixed for each line. They will never change.

$$y = \boxed{2}x + \boxed{1}$$

Every point on the line can be substituted for x and y; the equation with the values from these points will always be true

This equation is true if x = 0 and y = 1, or if x = 1 and y = 3, etc. (infinite pairs of solutions)

A. y = 3x - 1

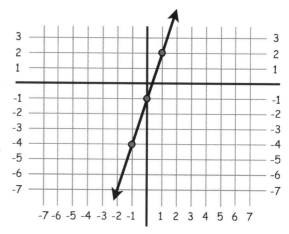

Three points on this line are shown. Plug them into the equation to show that POINTS ON A LINE make LINEAR EQUATIONS TRUE

Plug in the values for the point (-1,-4)

Plug in the values for the point (0,-1)

Plug in the values for the point (1,2)

( ) = 3( ) - 1   ( ) = 3( ) - 1   ( ) = 3( ) - 1

Did the slope change?   Did the y-intercept change?

42

1. $y = \frac{1}{2}x + 1$

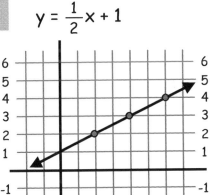

Plug in the points from line into the equation and show that they are valid solutions:

Point (2,2)

$(\quad) = \frac{1}{2}(\quad) + 1$

Point (4,3)

$(\quad) = \frac{1}{2}(\quad) + 1$

Point (6,4)

$(\quad) = \frac{1}{2}(\quad) + 1$

2. $y = 2x - 3$

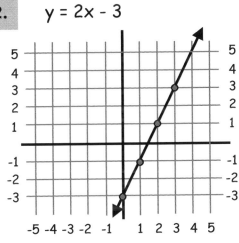

Plug in the points from line into the equation and show that they are valid solutions:

Point (1,-1)

$(\quad) = (\quad)(\quad) - 3$

Point (2,1)

$(\quad) = (\quad)(\quad) - 3$

Point (3,3)

$(\quad) = (\quad)(\quad) - 3$

3. $y = \frac{2}{3}x - 1$

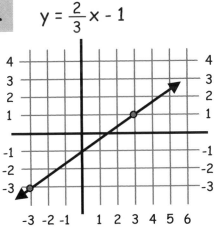

Plug in the points from line into the equation and show that they are valid solutions:

Point (0,-1)

$(\quad) = (\quad)(\quad) - 1$

Point (3,1)

$(\quad) = (\quad)(\quad) - 1$

How to Find a Missing y-Intercept

A. Find the y-intercept of a line that has a slope of 2 and the point (3,8)

- The slope is 2
- The point (3,8) is on the line

You know 3 out of 4 things. Your task is to find the value of "b"

$x = \boxed{3}$

$y = \boxed{8}$

THIS IS WHAT WE'RE LOOKING FOR!

$m = \boxed{2}$

$b = \boxed{}$

If you have any three of these values you can figure out the remaining one

$y = mx + \boxed{b}$

$8 = 2(3) + b$

$b = \boxed{2}$

---

1. A line has a slope of 3 and one point on the line is (4,13). What is the y-intercept?

$x = \boxed{}$ $m = \boxed{}$

$y = \boxed{}$ Plug in these values here

$y = mx + b$

$() = ()() + ()$

$b = \boxed{}$

---

2. A line has a slope of -2 and one point on the line is (3,2). What is the y-intercept?

$x = \boxed{}$ $m = \boxed{}$

$y = \boxed{}$

$() = ()() + ()$

$b = \boxed{}$

---

3. A line has a slope of 4 and one point on the line is (2,5). What is the y-intercept?

$x = \boxed{}$ $m = \boxed{}$

$y = \boxed{}$

$b = \boxed{}$

How to Find Any x- or y-Intercept

Look closely at the x- and y-axis. Find the zero next to the x-axis (representing y—because to be exactly on the x-axis the y-value has to be zero—not up or down any amount!).

Find the y-axis. Look at the zero at the bottom—otherwise the points would be off to the left or the right of the y-axis.

To find a y-intercept, make x = 0 and solve for y

To find an x-intercept, make y = 0 and solve for x

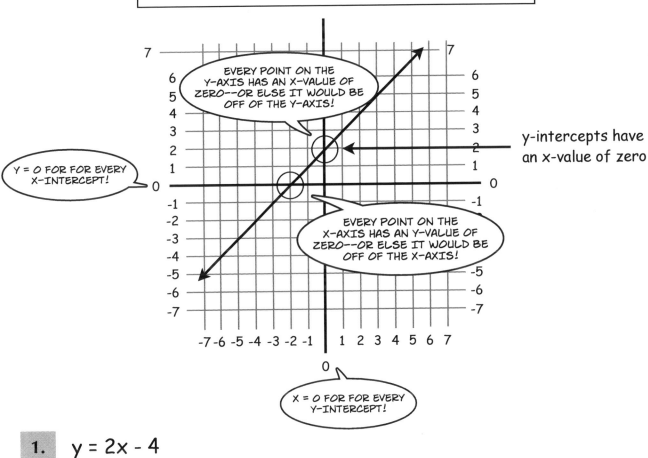

EVERY POINT ON THE Y-AXIS HAS AN X-VALUE OF ZERO—OR ELSE IT WOULD BE OFF OF THE Y-AXIS!

y-intercepts have an x-value of zero

Y = 0 FOR FOR EVERY X-INTERCEPT!

EVERY POINT ON THE X-AXIS HAS AN Y-VALUE OF ZERO—OR ELSE IT WOULD BE OFF OF THE X-AXIS!

X = 0 FOR FOR EVERY Y-INTERCEPT!

1. y = 2x - 4

MAKE Y ZERO TO FIND THE X-INTERCEPT!

MAKE X ZERO TO FIND THE Y-INTERCEPT!

( ) = 2x - 4   x-intercept = ☐   y = 2( ) - 4   y-intercept = ☐

© Peter Wise, 2015

45

Finding Intercepts

To find a y-intercept, make x = 0 and solve for y

To find a x-intercept, make y = 0 and solve for x

> TO FIND ANY INTERCEPT, MAKE THE OPPOSITE COORDINATE ZERO!

1. y = 4x - 8

> TO FIND THE X-INTERCEPT, PLUG IN ZERO FOR Y AND SOLVE!

> RECOMMENDATION: CHECK YOUR ANSWERS ON A GRAPH AFTERWARDS!

x-intercept =   ( ) = 4x - 8

☐ x-intercept   (☐ ,0) point

y-intercept =   y = 4( ) - 8

☐ y-intercept   (0,☐) point

> TO FIND THE Y-INTERCEPT, PLUG IN ZERO FOR X AND SOLVE!

Any intercept has to have one value be zero

Make the opposite coordinate zero and solve!

2. y = $\frac{1}{3}$x + 3

→ x-intercept =   ( ) = $\frac{1}{3}$x + 3

☐ x-intercept   ☐ point

→ y-intercept =   y = $\frac{1}{3}$( ) + 3

☐ y-intercept   ☐ point

3. y = $\frac{3}{4}$x - 6

> RECOMMENDATION: USE PARENTHESES FOR THE VARIABLE THAT YOU WILL SUBSTITUTE ZERO INTO!

x-intercept =

☐ x-intercept   ☐ point

y-intercept =

☐ y-intercept   ☐ point

4. y = -$\frac{1}{2}$x - $\frac{7}{2}$

x-intercept =

 x-intercept   point

y-intercept =

 y-intercept   point

Solving for Intercepts

Solve for intercepts by plugging in 0 for x or y

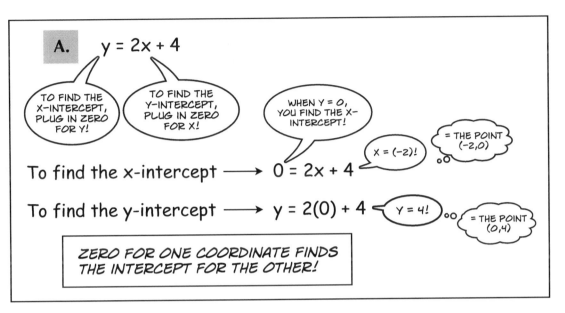

A. $y = 2x + 4$

TO FIND THE X-INTERCEPT, PLUG IN ZERO FOR Y!

TO FIND THE Y-INTERCEPT, PLUG IN ZERO FOR X!

WHEN Y = 0, YOU FIND THE X-INTERCEPT!

$x = (-2)!$

= THE POINT (-2,0)

To find the x-intercept $\longrightarrow$ $0 = 2x + 4$

To find the y-intercept $\longrightarrow$ $y = 2(0) + 4$ $\quad$ Y = 4!

= THE POINT (0,4)

ZERO FOR ONE COORDINATE FINDS THE INTERCEPT FOR THE OTHER!

1. $y = 6x + 12$ x-intercept ☐   y-intercept ☐

2. $y = 3x + 18$ x-intercept ☐   y-intercept ☐

3. $y = x + 5$ x-intercept ☐   y-intercept ☐

4. $y = -4x + 4$ x-intercept ☐   y-intercept ☐

5. $y = \frac{11}{3}x - 11$ x-intercept ☐   y-intercept ☐

Missing an x-value = parallel to x-axis

Missing a y-value = parallel to y-axis

A. Graph the line of the equation x = 5

RULE: If you are missing the y-value, the line will be parallel to the y-axis

Example

x	y
5	
5	
5	

THE X-VALUES ARE ONLY AND ALWAYS 5!

PUT ANY NUMBERS YOU WANT HERE!

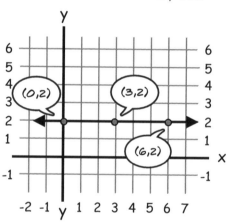

(5,6)

(5,1)

(5,0)

x = 5 — Don't see a y anywhere? The line will be parallel to the y-axis!

B. Graph the line of the equation y = 2

RULE: If you are missing the x-value, the line will be parallel to the x-axis

Example

Put in any numbers you want here!

x	y
→	2
→	2
→	2

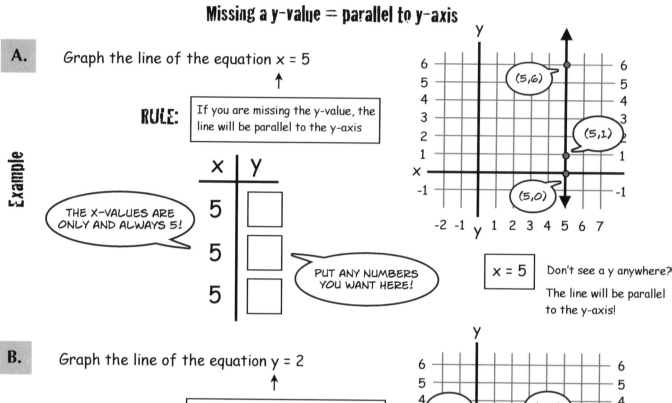

(0,2)   (3,2)

(6,2)

y = 2 — Don't see ax x anywhere! The line will be parallel to the x-axis!

1. Plug in values for the equation x = 6 and then graph the line

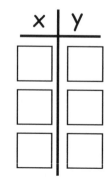

x	y

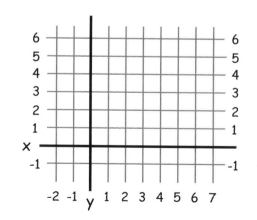

Practice with Missing x- or y-Values

They really aren't missing—you just don't always need to write them

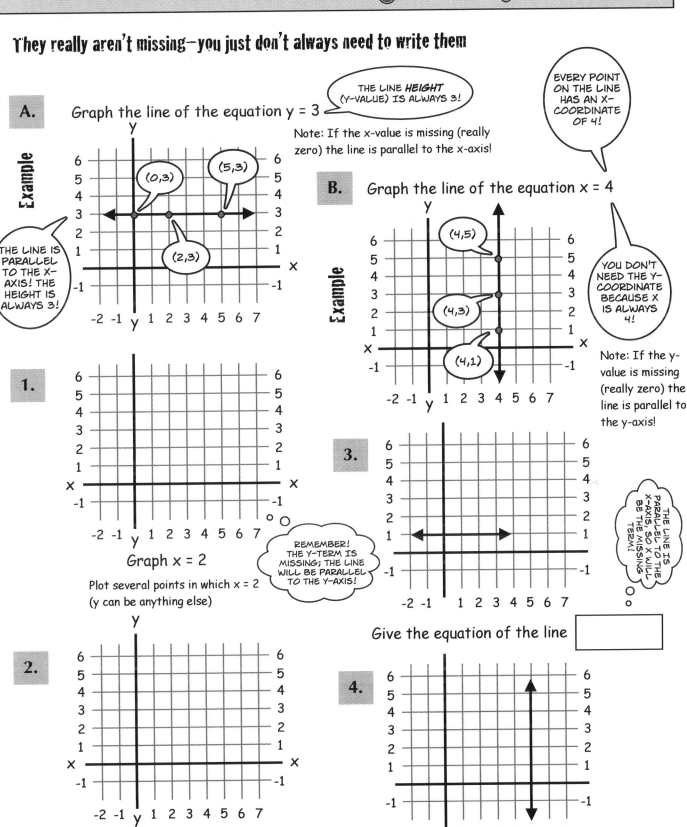

A. Graph the line of the equation y = 3

THE LINE *HEIGHT* (Y-VALUE) IS ALWAYS 3!

Note: If the x-value is missing (really zero) the line is parallel to the x-axis!

EVERY POINT ON THE LINE HAS AN X-COORDINATE OF 4!

Example

(0,3) (5,3) (2,3)

THE LINE IS PARALLEL TO THE X-AXIS! THE HEIGHT IS ALWAYS 3!

B. Graph the line of the equation x = 4

Example

(4,5) (4,3) (4,1)

YOU DON'T NEED THE Y-COORDINATE BECAUSE X IS ALWAYS 4!

Note: If the y-value is missing (really zero) the line is parallel to the y-axis!

1.

Graph x = 2

Plot several points in which x = 2 (y can be anything else)

REMEMBER! THE Y-TERM IS MISSING; THE LINE WILL BE PARALLEL TO THE Y-AXIS!

3.

THE LINE IS PARALLEL TO THE X-AXIS, SO X WILL BE THE MISSING TERM!

Give the equation of the line

2.

Graph y = 6

Plot several points in which y = 6 (x can be anything else)

4.

Give the equation of the line

© Peter Wise, 2015

49

Finding Slope from x/y Tables (Vertical)

1. Find the slope from the given x/y table

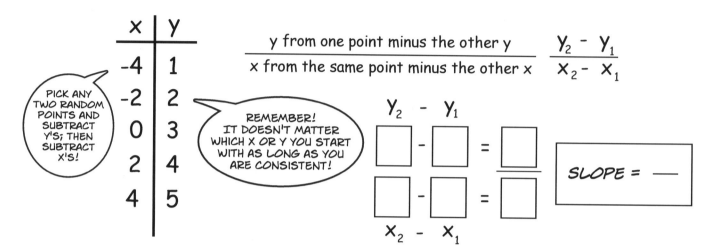

x	y
-4	1
-2	2
0	3
2	4
4	5

PICK ANY TWO RANDOM POINTS AND SUBTRACT Y'S; THEN SUBTRACT X'S!

REMEMBER! IT DOESN'T MATTER WHICH X OR Y YOU START WITH AS LONG AS YOU ARE CONSISTENT!

y from one point minus the other y / x from the same point minus the other x

$$\frac{Y_2 - Y_1}{X_2 - X_1}$$

$$\frac{Y_2 - Y_1}{\boxed{} - \boxed{}} = \frac{\boxed{}}{\boxed{}}$$

$$\boxed{} - \boxed{} = \frac{\boxed{}}{\boxed{}}$$

$X_2 - X_1$

SLOPE = —

2. Find the slope

NOW TRY OTHER POINTS!

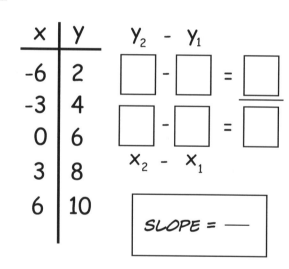

x	y
-6	2
-3	4
0	6
3	8
6	10

$Y_2 - Y_1$

$$\boxed{} - \boxed{} = \boxed{}$$

$$\boxed{} - \boxed{} = \boxed{}$$

$X_2 - X_1$

SLOPE = —

$Y_2 - Y_1$

$$\boxed{} - \boxed{} = \boxed{}$$

$$\boxed{} - \boxed{} = \boxed{}$$

$X_2 - X_1$

SLOPE = —

3.

x	y
-20	2
-15	4
-10	6
-5	8

$$\boxed{} - \boxed{} = \boxed{}$$

$$\boxed{} - \boxed{} = \boxed{}$$

SLOPE = —

4.

x	y
-20	2
-16	4
-12	6

$$\boxed{} - \boxed{} = \boxed{}$$

$$\boxed{} - \boxed{} = \boxed{}$$

SLOPE = —

Linear Equations Review

REMEMBER
THE GLASSES TRICK?

DRAW IN THE OVAL AND RECTANGLE FRAMES!

1. Find the slope of the line going through these points: (-10,-4), (-3,-1)

$$\frac{-}{-} = \frac{\boxed{}}{\boxed{}}$$

2. Plot three points (one x-value can be zero)

$$y = \frac{2}{3}x + 1$$

$$() = \frac{2}{3}() + 1$$

x	y

CHOOSE X-VALUES THAT ARE MULTIPLES OF 3 TO ELIMINATE FRACTIONS!

3. Slope is

$$\frac{\text{Change in } \boxed{}}{\text{Change in } \boxed{}}$$

4. Find the equation of the line with a slope of $\frac{3}{5}$ and going through the point (-5,1)

Remember to make a list of x, y, m, b

$$\boxed{}$$

5. Find the y-intercept (the b-value)

$$y = mx + b \qquad b = \boxed{}$$

$$() = ()() + ()$$

- The slope is $\frac{2}{7}$
- The point (21,2) is on the line

6. Find the x-intercept in the line $y = \frac{3}{4}x - 6$ x-intercept = $\boxed{}$

7. In the formula y = mx + b, which values are fixed numbers? $\boxed{}$

Which values are infinite pairs of solutions? $\boxed{}$

51

f(x) is read "f of x"

A. Linear equations like y = 2x + 1 are commonly written with f(x) instead of y in the equation.

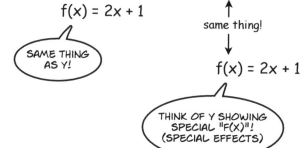

f(x) = 2x + 1

SAME THING AS Y!

y = 2x + 1

same thing!

f(x) = 2x + 1

THINK OF Y SHOWING SPECIAL "F(X)"! (SPECIAL EFFECTS)

f(x) is called a function; this will be covered in later pages!

...although you can also write this as g(x) or h(x), etc.

So don't have a heart attack if you see f(x). Just think of it as another way of writing y!

"f of x"

1. $f(x) = \frac{1}{3}x + 1$

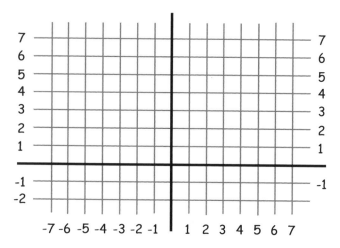

3. $f(x) = -\frac{5}{4}x + 4$

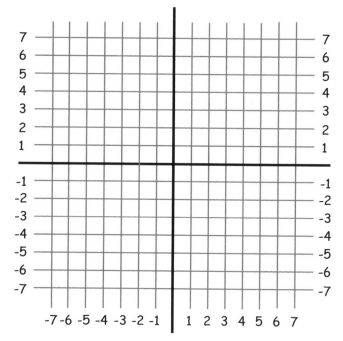

2. $f(x) = \frac{2}{3}x - 2$

Input multiples of 3 to cancel the fraction denominator!

x	3			
y				

Graphing Practice–Using f(x)

Graph the following lines

F(X) HERE IS THE SAME AS Y!

1. $f(x) = 3x - 7$

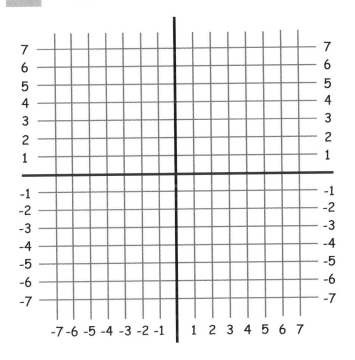

3. $f(x) = \frac{3}{4}x + 2$

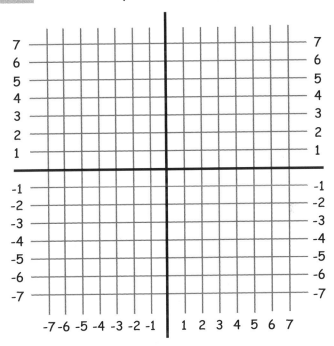

2. $f(x) = 2x - 5$

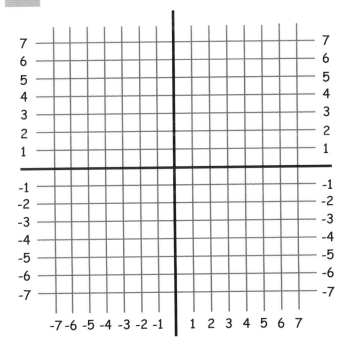

4. $f(x) = -\frac{3}{4}x + 6$

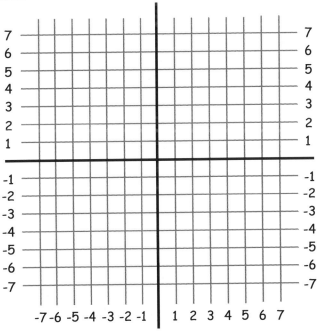

Concept Review

1. Which part(s) of y = mx + b have fixed numerical values?

   _____

2. Which part(s) of y = mx + b represent infinite pairs of solutions?

   _____

3. If you see the equation: y = 3 how would you write this using y = mx + b?

   _____

   What is the slope? _____

   Graph the line:

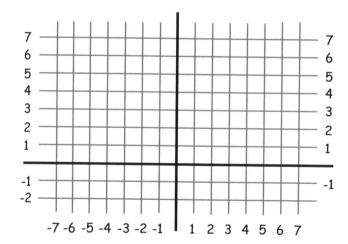

4. What are the two ways of plotting points with the formula y = mx + b?

   _____

   _____

5. f(x) is the same as: _____

   f(x) is read: _____

Introduction to Functions

Functions: Only ONE (unique) y-value for any x-value you input

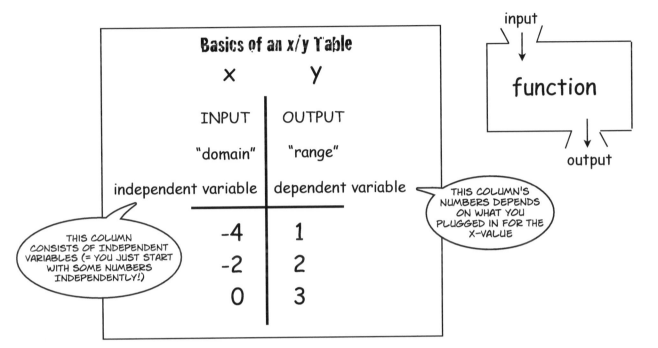

Basics of an x/y Table

x	y
INPUT	OUTPUT
"domain"	"range"
independent variable	dependent variable
-4	1
-2	2
0	3

THIS COLUMN CONSISTS OF INDEPENDENT VARIABLES (= YOU JUST START WITH SOME NUMBERS INDEPENDENTLY!)

THIS COLUMN'S NUMBERS DEPENDS ON WHAT YOU PLUGGED IN FOR THE X-VALUE

input → function → output

DEFINITION OF A FUNCTION: A relationship between a set of x- and y-values in which for every x-value, there is only ONE unique y-value

Just look for duplicate x-values. If the same x-values always give the same y-values, it is a function. (Ignore duplicate y-values.)

This is a Function:

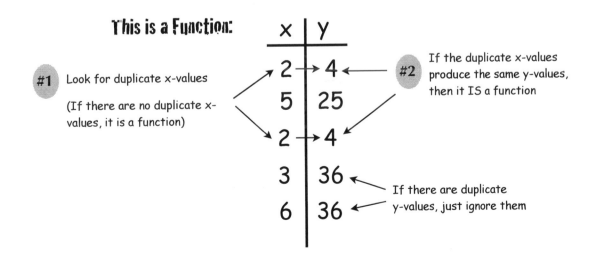

#1 Look for duplicate x-values

(If there are no duplicate x-values, it is a function)

x	y
2	4
5	25
2	4
3	36
6	36

#2 If the duplicate x-values produce the same y-values, then it IS a function

If there are duplicate y-values, just ignore them

Intro to Functions, pt. 2

Functions: Only ONE (unique) y-value for any x-value you input

Visual Characteristic: No two points on the graph of a function can line up vertically

Okay for a function:

DIFFERENT → SAME

X   Y

= two ways of getting the same (predictable) results

Not okay for a function:

SAME → DIFFERENT

X   Y

= unpredictable results

How to spot non-functions with tables:

Look for SD!
("same → different")

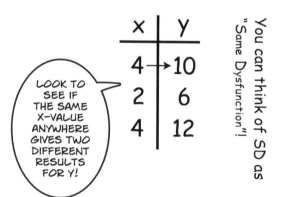

x	y
4 →	10
2	6
4	12

You can think of SD as "Same Dysfunction"!

LOOK TO SEE IF THE SAME X-VALUE ANYWHERE GIVES TWO DIFFERENT RESULTS FOR Y!

LOOK FOR DUPLICATE X-VALUES! IF THE SAME X-VALUES PRODUCE TWO DIFFERENT Y-VALUES, THEN THIS IS NOT A FUNCTION!

x	1	2	3	2
y	4	7	10	5

How to spot non-functions visually:

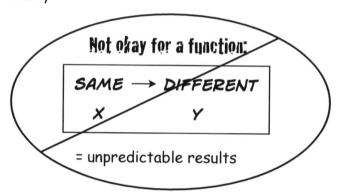

VERTICAL LINE TEST!

MANY SHAPES (LIKE CIRCLES) ARE VALID, BUT JUST AREN'T FUNCTIONS!

IF TWO POINTS LINE UP VERTICALLY, YOU KNOW THIS IS NOT A FUNCTION!

YOU INPUT 5 (X-VALUE), BUT GOT TWO RESULTS (2 AND 6)!

How to spot non-functions with lists of points:

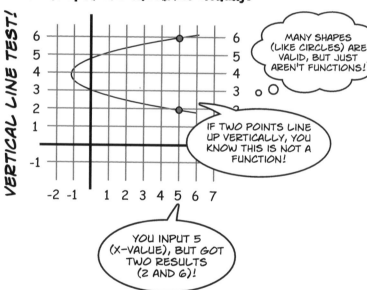

different results

(2,3)(4,5)(6,2)(7,8)(4,9)

same input

YOU HAVE TO HAVE DUPLICATE X-VALUES FOR A NON-FUNCTION!

#1 Look for two of the same x-values

#2 Check to see if the y-values are different. If they are, then it isn't a function! (unpredictable results)

Testing for Functions

#1 Look for duplicate x-values

(If there are no duplicate x-values, it is a function)

#2 If the duplicate x-values produce the SAME y-values, then it IS a function

If the duplicate x-values produce the DIFFERENT y-values, then it is NOT a function

If there are duplicate y-values, just ignore them

1.

x	y
3	5
8	20
2	7
7	8
8	20

function   not a function

circle one

2.

x	y
9	4
12	7
2	4
15	8
20	9

function   not a function

circle one

3.

x	8	14	7	8	15
y	5	8	4	3	9

function   not a function

4.

y	8	13	8	12	17
x	6	9	12	8	11

function   not a function

NOTICE THAT THE X– AND Y–VALUES ARE SWITCHED ON THIS PROBLEM!

5. (4,7) (5,8) (5,8) (10,14) (4,9) (6,11)

function   not a function

6.

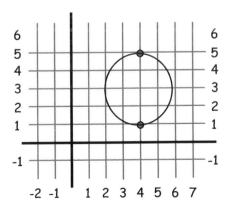

function   not a function

57

Testing for Functions

Tell if the following sets of points are functions

1.

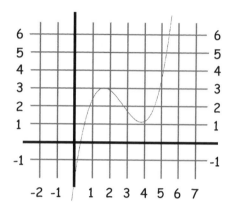

Yes   No

(circle one)

Careful! Duplicate y-values for different x-values are okay.

> ---JUST LIKE GETTING THE SAME ANSWER (Y-VALUE) TWO DIFFERENT WAYS!

5.   (3,5) (7,10) (3,6) (8,11)

Yes   No

2.

x	5	7	10	7
y	9	11	14	14

Yes   No

6.

x	y
-5	25
-2	4
1	1
-2	4

Yes   No

3.   (2,3) (4,6) (5,8) (7,3)

Yes   No

4.

x	y
-7	42
-5	21
-3	4
-1	-4
-3	-2

Yes   No

7.

x	-3	1	3	4
y	9	1	9	16

Yes   No

58

Finding the Value of One Coordinate

For all these use the formula: $y = mx + b$

If you are just missing one coordinate, plug in everything else you have and solve for the missing x- or y-value

TIP: There are four possible pieces of information (x,y,m,b)

If you are missing 1 value the other 3 will enable you to figure out the 4th!

1. Find the value of the y-coordinate if
   • the slope is 2
   • the y-intercept is 0
   • one point is (4,y)

   $y = ($ $)($ $) + ($ $)$

   DO YOU REMEMBER WHICH VALUE THE Y-INTERCEPT IS?

   x = ☐   m = ☐   b = ☐

   y = ☐

2. Find the value of the x-coordinate if
   • the slope is 3
   • the y-intercept is 2
   • one point is (x,8)

   y = ☐   m = ☐   b = ☐

   x = ☐

3. Find the value of the y-coordinate if
   • the slope is 2
   • the y-intercept is -1
   • one point is (4,y)

   x = ☐   m = ☐   b = ☐

   y = ☐

4. Find the value of the x-coordinate if
   • the slope is $\frac{2}{3}$
   • the y-intercept is 4
   • one point is (x,6)

   TO GET RID OF THE 2/3, MULTIPLY BOTH SIDES BY THE DENOMINATOR!

   y = ☐   m = ☐   b = ☐

   x = ☐

Finding Equations from 2 Points

Write equations from the two ordered pairs given

1. Find the equation of the line going through the points (1,5) and (2,9)

PICK ANY POINT YOU ARE GIVEN AND USE THE SAME X- AND Y-VALUES FOR THIS!

HOW MANY PIECES OF INFORMATION DO YOU HAVE FOR THE Y = MX + B FORMULA?

YOU HAVE AN X-VALUE, A Y-VALUE, AND SLOPE (AFTER YOU SUBTRACT THE Y-COORDINATES AND THE X-COORDINATES!)

x = ☐

y = ☐

m = ☐

b = ☐

SUBTRACT Y-VALUES (NUMERATOR) AND SUBTRACT THE X-VALUES (DENOMINATOR)!

ONCE YOU HAVE ALL THE OTHER VALUES, USE BASIC ALGEBRA TO SOLVE FOR "B"!

YOUR EQUATION WILL NOW NEED TO HAVE "X" AND "Y" AS VARIABLES AND "M" AND "B" AS NUMBERS!

solution:

$y = \boxed{}\ x + \boxed{}$

2. Find the equation of the line going through the points (2,3) and (4,4)

Pick either point to use it's x- and y-values (use both from the same point)

x = ☐

y = ☐

Subtract the y-values and x-values to find the slope:

m = ☐

Now use basic algebra to solve for b. This is usually the last variable you need to find.

b = ☐

solution:

$y = \boxed{}\ x + \boxed{}$

For the equation of the line you need to have "x" and "y" as variables and "m" and "b" as numbers ("constants")

Practice Finding Equations from 2 Points

1. Find the equation of the line going through the points (-1,6) and (-2,-9)

Pick either point to use it's x- and y-values (use both from the same point)

x = ☐

y = ☐

Subtract the y-values and x-values to find the slope:

m = ☐

Now use basic algebra to solve for b. This is usually the last variable you need to find.

b = ☐

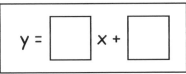

solution:

y = ☐ x + ☐

For the equation of the line you need to have "x" and "y" as variables and "m" and "b" as numbers ("constants")

2. Find the equation of the line going through the points (-4,-5) and (8,4)

x = ☐

y = ☐

m = ☐

b = ☐

solution:

y = ☐ x ☐ ☐

3. Find the equation of the line going through the points (7,-7) and (-7,1)

x = ☐

y = ☐

m = ☐

b = ☐

solution:

y = ☐ x ☐ ☐

1. Find the equation of the line going through the points (3,4) and (-6,-2).

x = ☐

y = ☐

m = ☐

b = ☐

solution:

y = ☐ × ☐ ☐

2. Find the equation of the line going through the points (-3,-8) and (4,6).

x = ☐

y = ☐

m = ☐

b = ☐

solution:

y =

3. Find the equation of the line going through the points (-9,-2) and (9,-12)

x = ☐

y = ☐

m = ☐

b = ☐

solution:

Finding Slope from Rate of Change

TIME	DISTANCE
x value	y value
1	5
2	10

How to figure out slope from a chart

1. Treat the values like x and y values from a table

2. The rate of change is just like finding slope from two points

3. Subtract y's (numerator) and x's (denominator)

4. Your result is the slope!

REMEMBER! ONLY Y-VALUES GO HERE ON TOP!

THESE "RATE OF CHANGE" PROBLEMS ARE JUST LIKE FINDING SLOPE FROM TWO POINTS!

SLOPE = —

SLOPE =
(rate of change)

1.

TIME (HRS)	MONEY ($)
x value	y value
3	45
4	60

$$\frac{\bigcirc - \bigcirc}{\square - \square} = \frac{\square}{\square}$$ dollars / hour

2.

TIME (SEC)	HEIGHT (FT)
x value	y value
5	200
7	280

$$\frac{\bigcirc - \bigcirc}{\square - \square} = \frac{\square}{\square} = \frac{\square}{\square}$$ feet / sec

(simplify)

When Points Represent Values

1.

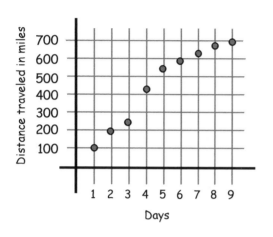

Days

Find the average number of miles traveled from day 2 to day 6

#1 Find the points on the lines for day 2 and day 6 (circle these points)

#2 Find the distance traveled for days 2 and 6

#3 Subtract these values and put your answer as the numerator (just like the "change in y")

#4 Now subtract the years and put your answer as the denominator (just like the "change in x")

THE ORDER IN WHICH YOU SUBTRACT DOESN'T MATTER; JUST MAKE SURE IT'S CONSISTENT!

Subtract the **vertical values** (like the y-values in regular graphs):

Subtract the **horizontal values** (like the y-values in regular graphs):

$$\frac{\text{Difference in miles traveled}}{\text{Difference in days}} = \frac{\boxed{} \text{ miles}}{\boxed{} \text{ days}} = \boxed{} \text{ miles per day}$$

2.

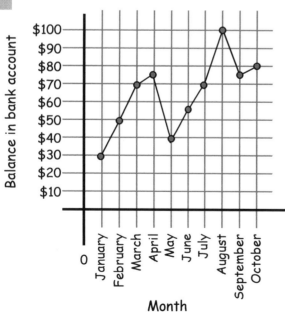

Month

Find the average change per month in this person's bank account from May to August

$$\frac{\text{Difference in dollar amounts}}{\text{Difference in months}} = \frac{\boxed{} \text{ dollars}}{\boxed{} \text{ months}}$$

$$= \boxed{} \text{ dollars per month}$$

When Points Represent Values

1.

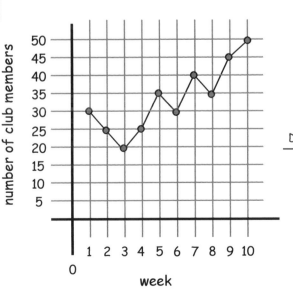

(Graph: Total website hits vs. years 2000–2008)

Find the rate of change (just like slope) from 2002 and 2006

THE ORDER IN WHICH YOU SUBTRACT DOESN'T MATTER; JUST MAKE SURE IT'S CONSISTENT!

#1 Find the points on the lines for 2002 and 2006 (circle these years)

#2 Find the values for "Total website hits" for years 2002 and 2006

#3 Subtract these values and put your answer as the numerator (just like the "change in y")

#4 Now subtract the years and put your answer as the denominator (just like the "change in x")

Subtract the **vertical values** (like the y-values in regular graphs):

Subtract the **horizontal values** (like the y-values in regular graphs):

$$\frac{\text{Difference in website hits}}{\text{Difference in years}} = \frac{\boxed{} \text{ hits}}{\boxed{} \text{ years}} = \boxed{} \text{ hits per year}$$

2.

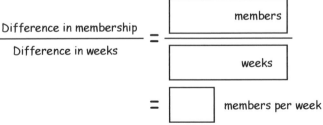

(Graph: number of club members vs. week 1–10)

Find the rate of change (slope) between week 4 and week 9

$$\frac{\text{Difference in membership}}{\text{Difference in weeks}} = \frac{\boxed{} \text{ members}}{\boxed{} \text{ weeks}}$$

$$= \boxed{} \text{ members per week}$$

1. Give the meaning of these components of y = mx + b

m = _____

b = _____

_____

2.

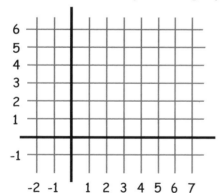

y-intercept ☐   Slope = $\dfrac{\square}{\square}$

4. Find the y-intercept

PLUG IN EVERY PIECE OF INFORMATION YOU HAVE!

- The slope is 1/4
- The point (8,6) is on the line

y = mx + b   b = ☐

3. Make a table and plot the graph

$y = -\dfrac{1}{5}x + 4$

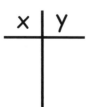

5. Find the equation of the line with a slope of 2/7 and going through the point (14,12)

HINT THINK ABOUT THE PIECE YOU ARE MISSING!

1.

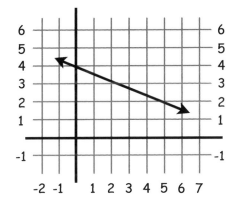

y-intercept ☐   Slope = $\dfrac{☐}{☐}$

Give the equation of the line:

y = ☐ x ☐ ☐

2. Make a table and plot the graph

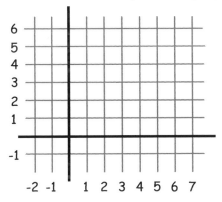

$y = \dfrac{3}{4}x - 1$

x	y

USE MULTIPLES OF 4 TO CANCEL OUT THE 4 IN THE DENOMINATOR!

3. Find the y-intercept

- The slope is 1/3
- The point (9,1) is on the line

y = mx + b   b = ☐

4. Find the value of the x-coordinate if
- the slope is 4
- the y-intercept is 3
- one point is (x,11)   x =

5. Find the slope of the line going through these two points:

(2,6) (7,5)

6. Find the equation of the line with a slope of 2/3 and going through the point (9,11)

67

Concept Quiz

1. Slope is defined as "change in

[ ]

2. y = 2x + b has how many solution pairs for x and y?

[ ]

3. If you have a y-intercept of 4, the x-value of that point is

[ ]

4. If you have an x-intercept of 2, the y-value of that point is

[ ]

5. When you have an equation like x = 3 the line is   **vertical   horizontal**

(circle one)

Rule with missing x or y values:

[ ]

6. When you calculate slope from two points with $y_2 - y_1$, how do you know which y is y_2 and which y is y_1?

[ ]

7. If you have two points: (5,7) and (10,10), find each of the following:

(a) slope
(b) y-intercept
(c) the equation of the line

[ ]

Linear Equations Quiz

1.

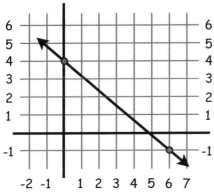

y-intercept ☐   Slope = $\frac{\square}{\square}$

Give the equation of the line:

y = ☐ x ☐ ☐

2. Make a table and plot the graph

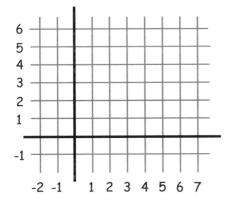

$y = \frac{2}{5}x - 1$

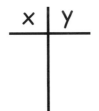

x	y

3. Find the y-intercept

- The slope is 2/3
- The point (12,1) is on the line

b = ☐

4. Find the value of the x-coordinate if
- the slope is 3
- the y-intercept is 6
- one point is (x,-6)

5. Find the slope of the line going through these two points:

(3,7) (5,10)

6. Find the equation of the line with a slope of 2/5 and going through the point (5,9)

7. Give the equation for the line that goes through the points (14,5) and (7,7)

Standard Form Equations

Standard Form Equations:

$$Ax + By = C$$

A, B, and C are constants (numbers as opposed to letters)

Standard form equations have x and y on the left side of the equation

Convert the Standard Form equation to Slope-Intercept Form (y = mx + b)

Standard Form Equation

A. $6x + 3y = 12$

↑   ↑
Both variables are
on the left side

$6x + 3y = 12$

$3y = -6x + 12$ ← Subtract 6x from both sides

$y = -2x + 4$ ← Divide every term by 3

Slope-Intercept Form Equation

Convert the Standard Form equations to Slope-Intercept (y = mx + b) equations

1. $-2x + y = 5$

← Move the term with x to the other side by adding 2x to both sides

$y = $ ☐ ← Put the term with x first

Slope-Intercept Form

3. $-11x + 2y = 9$

← Get the x-term on the right side

☐ ← Divide both sides by 2; some terms will be improper fractions

Slope-Intercept Form

2. $-7x + y = 14$

← Get the x-term to the right side of the equation

$y = $ ☐

4. $-\frac{1}{2}x + \frac{5}{3}y = 20$

← Move the -1/2x to the other side by adding -1/2x to both sides

☐ ← Get rid of the fraction coefficient of y by multiplying every term by 3/5

Intro to Point-Slope Form

Point-Slope form is just another way of writing y = mx + b

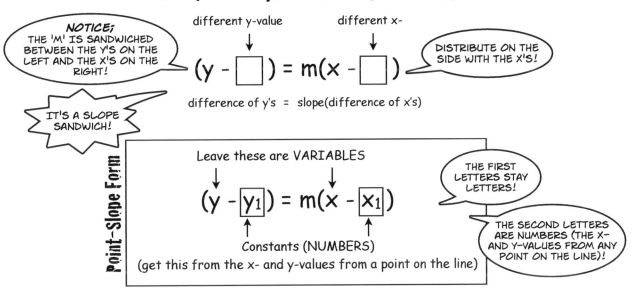

NOTICE; THE 'M' IS SANDWICHED BETWEEN THE Y'S ON THE LEFT AND THE X'S ON THE RIGHT!

IT'S A SLOPE SANDWICH!

different y-value

different x-

$$(y - \boxed{}) = m(x - \boxed{})$$

difference of y's = slope(difference of x's)

DISTRIBUTE ON THE SIDE WITH THE X'S!

Point-Slope Form

Leave these are VARIABLES

$$(y - \boxed{y_1}) = m(x - \boxed{x_1})$$

Constants (NUMBERS)
(get this from the x- and y-values from a point on the line)

THE FIRST LETTERS STAY LETTERS!

THE SECOND LETTERS ARE NUMBERS (THE X- AND Y-VALUES FROM ANY POINT ON THE LINE)!

Example:

A. Put into Point-Slope form: the point (2,1) with the slope (m) = 3

$$(y - \boxed{1}) = \boxed{3}(x - \boxed{2})$$

Write POINT-SLOPE Equations using the point and slope you are given

1. Put into Point-Slope form: the point (9,2) with the slope (m) = 6

$$(y - \boxed{}) = \boxed{}(x - \boxed{})$$

difference of y's = slope(difference of x's)

2. Put into Point-Slope form: the point (1,5) with the slope of -4

copy → $(y - \boxed{}) = \boxed{}(x - \boxed{})$ ← distribute the right side

$(\boxed{}) = (\boxed{})$

3. Put into Point-Slope form: the point (7,3) with the slope (m) = 2

$(\boxed{}) = \boxed{}(\boxed{})$ → copy and distribute → $(\boxed{}) = (\boxed{})$

71

More on Point-Slope Form

SLOPE of a line

$$\frac{(y - y_1)}{(x - x_1)} = m$$

Multiply both sides by $(x - x_1)$

$$(x - x_1)\frac{(y - y_1)}{(x - x_1)} = m(x - x_1)$$

these cancel out

Point-Slope form

$$(y - y_1) = m(x - x_1)$$

You can rewrite this into Slope-Intercept $(y = mx + b)$ form using basic algebra

---

Put into POINT-SLOPE form, and then convert to SLOPE-INTERCEPT form

1. Point (4,2); slope of 3

$$(y - \boxed{}) = \boxed{}(x - \boxed{})$$ ← distribute the right side

copy: → $$(\boxed{}) = (\boxed{})$$

add 2 to both sides: → $$\boxed{} = $$

NOW YOU HAVE THE EQUATION BACK IN Y = MX + B FORM!

...AKA SLOPE-INTERCEPT FORM!

2. Point (6,1); slope of $\frac{1}{2}$

$$(y - \boxed{}) = \boxed{}(x - \boxed{})$$ ← distribute the right side

$$(\boxed{}) = (\boxed{})$$

add 1 to both sides: → $$\boxed{} = $$

3. Point (2,3); slope of 4

$$\boxed{} = \boxed{}$$ ← distribute

$$\boxed{} = \boxed{}$$

Get the y-term by itself: → $$\boxed{} = $$

Graphing Point-Slope Form

Convert to SLOPE-INTERCEPT form, then graph

1. $y - 1 = 2(x - 4)$

([   ]) = ([   ])

[   ] =

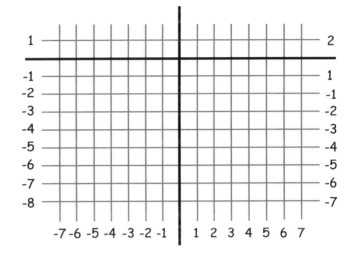

2. $y - (-1) = -\frac{7}{4}(x - 4)$

[   ] = [   ]

final equation in y = mx + b form:

[   ] = [   ]

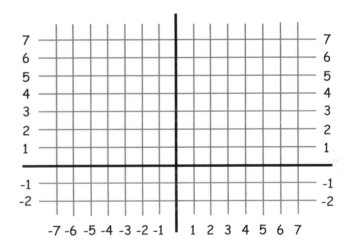

3. Point (-5,-3); slope of $-\frac{2}{5}$

put into point-slope form:

 = [   ]

simplify:

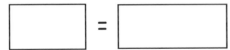 = [   ]

final equation in y = mx + b form:

[   ]

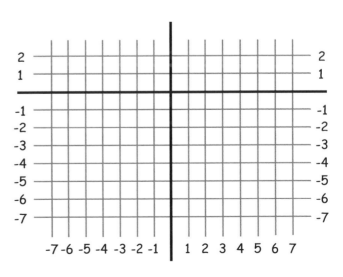

73

Converting Point-Slope to Slope-Intercept

Using basic algebra, you can convert Point-Slope Equations to Slope-Intercept Equations

$$(y - y_1) = m(x - x_1) \qquad y = mx + b$$

A.

$y - 5 = 2(x - 1)$ ← Distribute the 2

$y - 5 = 2x - 2$ ← Add 5 to both sides

$y = 2x + 3$

Convert the Point-Slope equations to Slope-Intercept (y = mx + b) equations

---

1. $y - 2 = 5(x + 3)$

[ ] ← Distribute the 5

[ ] ← Add 2 to both sides

y = mx + b form

4. $y + 4 = -\dfrac{4}{5}(x + 5)$

[ ] ← Distribute

[ ] ← Solve for y

y = mx + b form

---

2. $y - 8 = \dfrac{1}{2}(x - 4)$

[ ] ← Distribute

[ ] ← Isolate the y-variable

y = mx + b form

NOTE: ANSWER WILL HAVE FRACTONS!

5. $y - (-3) = \dfrac{5}{7}[x - (-2)]$

[ ] ← Distribute

[ ] ← Solve for y

y = mx + b form

SAME WITH THIS CHALLENGE PROBLEM!

3. $y - 6 = -3(x - 9)$

[ ] ← Distribute

[ ] ← Isolate the y-variable

y = mx + b form

6. $y - (-4) = -\dfrac{2}{5}[x - (-3)]$

[ ] ← Distribute

[ ] ← Solve for y

y = mx + b form

Review of Standard Form and Point-Slope Form

1. Convert the following equation from Standard Form to SLOPE-INTERCEPT form:

   $3x + y = 8$ $\longrightarrow$ [ ]

2. Convert the following equation from Slope-Intercept to STANDARD form:

   $y = 7x + 5$ $\longrightarrow$ [ ]

3. Convert the following equation to SLOPE-INTERCEPT form:

   $6y = 18x + 12$ $\longrightarrow$ [ ]

4. Convert the following equation to SLOPE-INTERCEPT form:

   $-11x + 13y = 15$ $\longrightarrow$ [ ]

5. Point (32,-1); slope of $-\frac{3}{8}$

   put into point-slope form:

   [ ] = [ ]

   simplify:

   [ ] = [ ]

   final equation in $y = mx + b$ form:

   [ ]

6. Point (10,-2); slope of $\frac{4}{5}$

   put into point-slope form:

   [ ] = [ ]

   simplify:

   [ ] = [ ]

   final equation in $y = mx + b$ form:

   [ ]

Review: Finding Equations from 2 Points

1. Find the equation of the line going through the points (1,8) and (3,14)

x = ☐

y = ☐

m = ☐

b = ☐

solution:

y = ☐ x + ☐

2. Find the equation of the line going through the points (2,-8) and (3,-10)

x = ☐

y = ☐

m = ☐

b = ☐

solution:

y =

3. Find the equation of the line going through the points (6,3) and (9,5)

x = ☐

y = ☐

m = ☐

b = ☐

solution:

Introduction to Linear Inequalities

Inequality basics

1. CHANGE THE INEQUALITY SIGN to an EQUAL SIGN and solve the linear equation normally (y = mx + b)

    a) This line is called the **"boundary"**

2. The kind of line you graph depends on the type of inequality

    a) If the inequality sign is ≤ or ≥ the boundary line will be SOLID

    b) If the inequality sign is < or > the boundary line will be DASHED
    (to show that "y" DOESN'T INCLUDE the boundary line—
    y is GREATER THAN or LESS THAN this boundary line)

3. If "y" is **GREATER THAN**, shade **ABOVE** the boundary line;
   if "y" is **LESS THAN**, shade the **BELOW** the boundary line.

Example:

A.

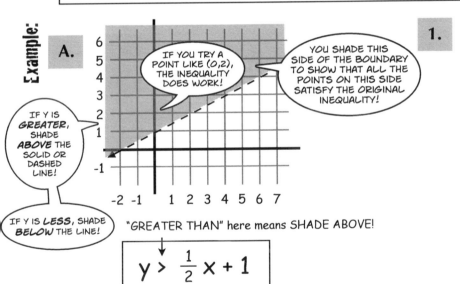

IF YOU TRY A POINT LIKE (0,2), THE INEQUALITY DOES WORK!

YOU SHADE THIS SIDE OF THE BOUNDARY TO SHOW THAT ALL THE POINTS ON THIS SIDE SATISFY THE ORIGINAL INEQUALITY!

IF Y IS **GREATER**, SHADE **ABOVE** THE SOLID OR DASHED LINE!

IF Y IS **LESS**, SHADE **BELOW** THE LINE!

"GREATER THAN" here means SHADE ABOVE!

$$y > \frac{1}{2}x + 1$$

- Graph the line as if it were y = 1/2x + 1
- Since y is **greater** (doesn't equal in the original inequality), change the line to a **dashed** boundary
- Since y is GREATER THAN the other part of the inequality, shade ABOVE the boundary line

1. Draw the correct boundary line and shade above or below

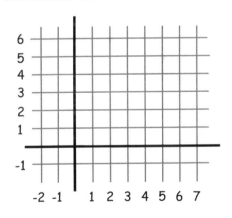

$$y > 2x - 1$$

TESTER POINT TRICK: Plug in any point above or below the line; if the values of a point above the line makes the inequality true shade in that region; if a point below works, shade below.

77

Linear Inequalities, pt. 2

Inequalities with ≥ or ≤

1. Make a boundary line by changing the inequality sign to an equal sign and solving the linear equation normally (y = mx + b)

2. Since ≥ means both > and =, the graph will have a **solid** line (which represents the = part) and be shaded to one side of the line (which represents the > part)

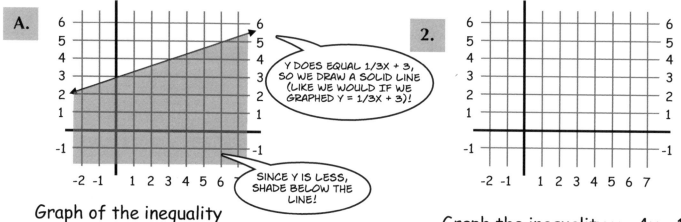

A.

Y DOES EQUAL 1/3X + 3, SO WE DRAW A SOLID LINE (LIKE WE WOULD IF WE GRAPHED Y = 1/3X + 3)!

SINCE Y IS LESS, SHADE BELOW THE LINE!

Graph of the inequality
y = ≤ 1/3x + 3

2.

Graph the inequality y ≤ 4x - 1

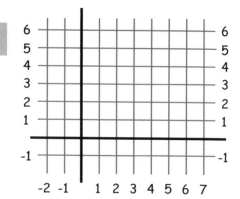

1.

Graph the inequality y ≥ 3x + 1

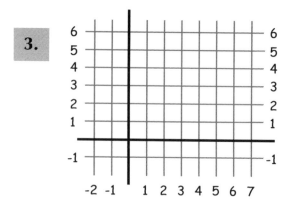

3.

Graph the inequality y ≥ -2/3x + 3

Change the ≥ to a = to make a boundary line;
if y is greater, shade above the line, if y is less, shade below the line.

Linear Inequalities Practice

1.

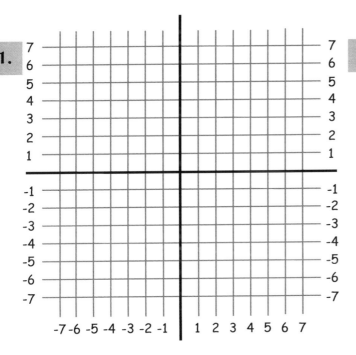

Graph the inequality y < 4x - 5

3.

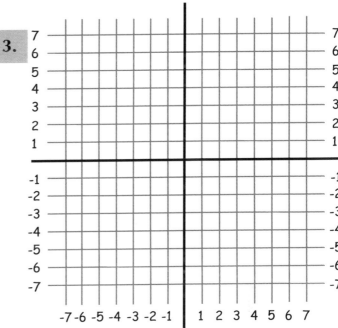

Graph the inequality y ≥ -3x + 6

2.

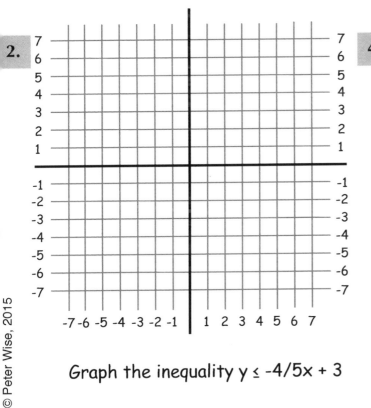

Graph the inequality y ≤ -4/5x + 3

4.

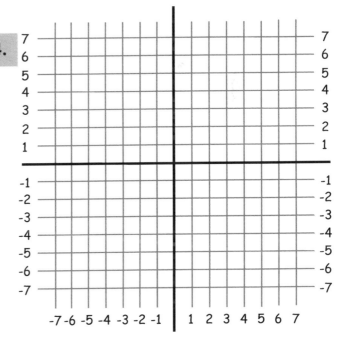

Graph the inequality y > 3/4x - 4

Review

1a.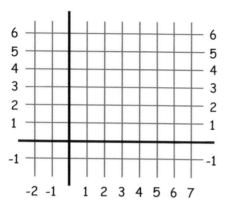

Graph y = 2 *THIS IS REALLY THE SAME AS Y = (0)X + 2*

1b.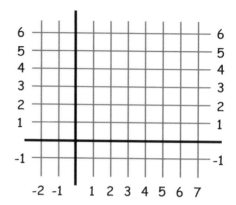

Graph y ≥ 2 *THIS IS THE SAME AS BOTH Y = 2 AND Y > 2!*

2a.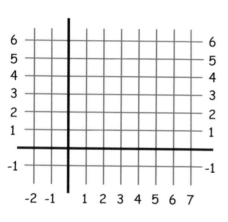

Graph x = 3

2b.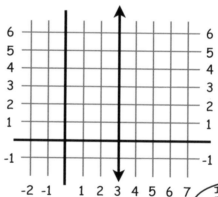

Graph x ≥ 3

IF X IS LESS THAN, SHADE TO THE LEFT!

IF X IS GREATER THAN THE OTHER VALUES, SHADE TO THE RIGHT!

The slope of this kind of line is _____

PLOT TWO POINTS ON THE LINE AND TRY FIGURING OUT THE SLOPE!

Review

3.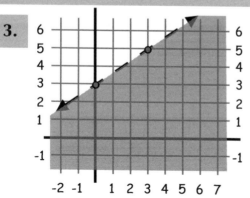

Write the inequality for this graph

<table><tr><td>

</td></tr></table>

4. Find the y-intercept   b = [ ]

- The slope is 2/5
- The point (5,9) is on the line

Review

1. Find the slope of the line going through these two points:

(3,2) (6,8)

5. Find the equation of the line having a slope of $\frac{3}{7}$ going through the point (7,12)

2. Find the y-intercept if

- the slope is $\frac{4}{5}$
- the point (5,6) is on the line

$y = mx + b$ $b =$ [   ]

6.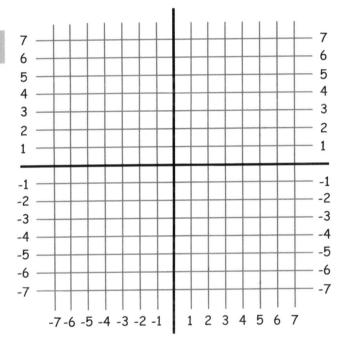

Graph the inequality $y \le -5/3x + 6$

3.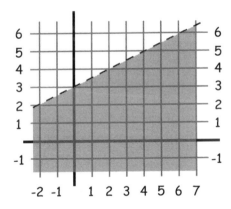

Write the inequality for this graph

7. Find the equation of the line going through the points (3,3) and (6,5)

4. Find the value of the x-coordinate if

- the slope is $\frac{4}{3}$
- the y-intercept is -3
- one point is (x,5)

$x =$

Advanced Material: Systems of Equations

- Solving by Graphing

- Solving by Elimination

- Solving by Substitution

YOUR TASK WITH SYSTEMS OF EQUATIONS:

Find values that make all the equations true at the same time

1.

$y = \frac{1}{2}x - 1$

$y = -2x + 4$

Find the point where both lines intersect.
The x- and y-values of this point are solutions for BOTH EQUATIONS!

THE X- AND Y-VALUES OF THIS POINT ARE SOLUTIONS FOR BOTH EQUATIONS!

ALL THE X- AND Y-VALUES FOR THE POINTS ON THIS LINE ARE SOLUTIONS OF $y = 1/2x - 1$!

ALL THE X- AND Y-VALUES FOR THE POINTS ON THIS LINE ARE SOLUTIONS OF $y = -2x + 4$!

Prove it

Plug in the x- and y-values of the INTERSECTION POINT into both equations and check to see that they are both valid!

INTERSECTION POINT: (,   )

$y = \frac{1}{2}x - 1$

$(\quad) = \frac{1}{2}(\quad) - 1$

y-value of the intersection point

x-value of the intersection point

$y = -2x + 4$

$(\quad) = -2(\quad) + 4$

2.

$y = x + 2$

$y = 5x - 2$

The INTERSECTION POINT contains the x- and y-values that solve BOTH EQUATIONS

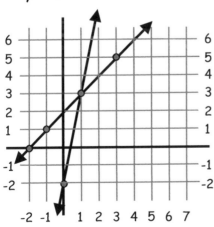

Plug in the x- and y-values of this point into both equations and check to see that they are both valid!

INTERSECTION POINT: (,   )

$y = x + 2$

$y = 5x - 2$

A. Find the solution for the following system of equations:

Example

a. y = x - 3

b. y = -4x + 2

Y = -4X + 2

ALL THE SOLUTIONS FOR THIS EQUATION ARE ON THIS LINE!

Y = X - 3

ALL THE SOLUTIONS FOR THIS EQUATION ARE ON THIS LINE!

THE SOLUTION FOR BOTH EQUATIONS IS ONLY THE POINT (1, -2)!

STEPS:

#1 Graph the line for each equation

#2 Find the point of intersection

THE SOLUTION TO BOTH EQUATIONS IS THIS INTERSECTION POINT!

DO YOU GET CORRECT ANSWERS FOR BOTH EQUATIONS?

Check It Out

PLUG IN THE POINTS (1, -2)

a. y = x - 3

(   ) = (   ) - 3

b. y = -4x + 2

(   ) = -4(   ) + 2

THE POINT OF INTERSECTION IS THE SOLUTION TO A SYSTEM OF EQUATIONS!

Graph each line find where the lines intersect

1. Find the solution for the following system of equations:

a. Graph y = 2x + 3

b. Graph y = -2x + 7

c. Find the point where the lines intersect

Solution: (,   )

= Point of Intersection

= Solution to the system of equations

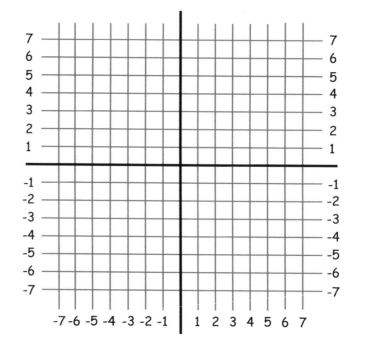

Systems of Equations: Solving by Graphing

DEFINITION: Finding values that make two or more equations true at the same time

1. Find the solution for the following system of equations:

a. $y = \frac{3}{2}x + 3$

b. $y = -\frac{1}{2}x - 1$

Solution: (,)
= Point of Intersection

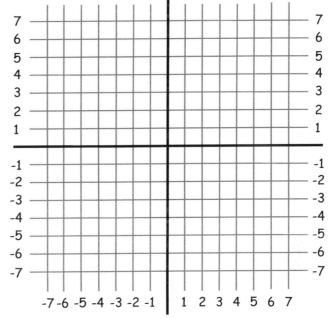

Plug in to check:

a. $y = \frac{3}{2}x + 3$

( ) = $\frac{3}{2}$( ) + 3

b. $y = -\frac{1}{2}x - 1$

( ) = $-\frac{1}{2}$( ) - 1

---

2. Find the solution for the following system of equations:

a. $y = -2x - 4$

b. $y = 4x + 2$

Solution: (,)
= Point of Intersection

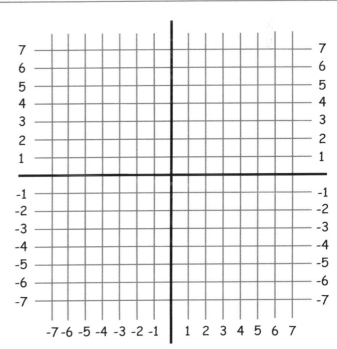

Plug in to check:

a. $y = -2x - 4$

( ) = -2( ) - 4

b. $y = 4x + 2$

( ) = 4( ) + 2

1. Find the solution for the following system of equations:

$$y = \frac{2}{3}x + 3$$

$$y = -\frac{1}{3}x$$

WHAT IS THE Y-INTERCEPT IF IT'S MISSING?

Solution: (,)

= Point of Intersection

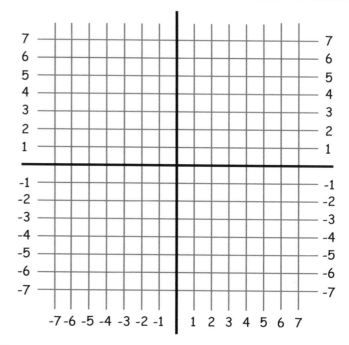

Plug in to check:

a. $y = \frac{2}{3}x + 3$

b. $y = -\frac{1}{3}x$

---

2. Find the solution for the following system of equations:

$$y = 2x - 3$$

$$y = -\frac{1}{2}x + 7$$

Solution: (,)

= Point of Intersection

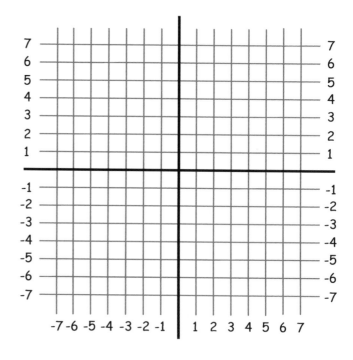

Plug in to check:

a. $y = 2x - 3$

b. $y = -\frac{1}{2}x + 7$

Systems of Equations: Solving by Elimination

EQUATIONS CAN BE ADDED

A.

$$x + 3 = 10$$
$$+ \quad x + 3 = 10$$
$$\overline{2x + 6 = 20}$$

THIS IS REALLY THE SAME AS MULTIPLYING BOTH SIDES OF AN EQUATION BY TWO!

Steps:

#1 Add the two equations to cancel out one variable and to figure out the the other variable

#2 Plug in the value of the variable you know to figure out the value of the one you don't know

Now you know both variables!

How to Figure Out One Variable at a Time

ELIMINATION METHOD

The y's cancel out so now you can figure out the value of x!

B.

$$x + y = 10$$
$$+ \quad x - y = 6$$
$$\overline{2x \quad = 16}$$
$$\boxed{x \quad = 8}$$

$$8 + y = 10$$
$$\boxed{y = 2}$$

$$\boxed{x = 8 \quad y = 2}$$

SOLVE THE FOLLOWING SYSTEMS OF EQUATIONS USING THE ELIMINATION METHOD

1.

$$x + y = 8$$
$$+ \quad x - y = 4$$

#1 Add the two equations to cancel out one variable and to figure out the the other variable

LOOK FOR THE EASIEST VARIABLE TO ELIMINATE!

#2 Plug in the value of the x into either equation to figure out y

x =   y =

LOOK FOR THE SIMPLEST EQUATION!

2.

$$3x + 2y = 8$$
$$+ \quad 3x - 2y = 4$$

#1 Add the two equations to cancel out one variable and to figure out the the other variable

#2 Plug in the value of the x into either equation to figure out y

x =   y =

© Peter Wise, 2015

87

Solve the following systems of equations using the ELIMINATION method

1.

$$+\begin{array}{l} 10x + 4y = -12 \\ -10x - 5y = 10 \end{array}$$

[ ]

#1 Add the two equations to cancel out one variable and to figure out the the other variable

[ ]

#2 Plug in the value of the y into either equation to figure out x

x = [ ]   y = [ ]

2.

$$+\begin{array}{l} 3x + 5y = 23 \\ -2x - 5y = -22 \end{array}$$

[ ]

#1 Add and solve for x

[ ]

#2 Plug in the value of the x to figure out y

x = [ ]   y = [ ]

3.

$$+\begin{array}{l} -7x + 3y = 13 \\ 8x - 3y = -14 \end{array}$$

[ ]

#1 Solve for x

[ ]

#2 Solve for y

x = [ ]   y = [ ]

4.

$$+\begin{array}{l} -5x + 3y = 3 \\ 5x + 6y = 6 \end{array}$$

[ ]

#1 Solve for y

[ ]

#2 Solve for x

x = [ ]   y = [ ]

Systems of Equations; Solving by Substitution

A.

$2x + 3y = 24$

$y = 2x$

SWAP THE Y FOR 2X!

$2x + 3(2x) = 24$

THIS IS WHAT YOU WANT... ALL X'S OR Y'S IN THE EQUATION!

$2x + 6x = 24$

$8x = 24$ $x = 3$

Steps:

#1 Look for x = some value or y = some value

#2 Plug in the value of the x or y into either equation to figure out the other letter

#3 Once you figure out one letter, plug in that value into either equation to figure out the other letter

BUT WHAT ABOUT Y?

Plug in x in either equation and solve for y!

$y = 2x$
$y = 2(3)$
$y = 6$

1.

$4x + 2y = 40$

$y = 8x$

SOLVE FOR X:

← Substitute 8x for y in the equation above. Write the (8x) in parentheses

← Multiply. Copy every other term in the equation.

← Combine terms. Divide by the coefficient of x.

$x = \boxed{}$

SOLVE FOR Y:

← Plug in your value for x into either equation. The second equation is simpler, so let's use that one. Don't forget to put the value for x in parentheses.

Point where the two lines cross (solution to both equations):

$y = \boxed{}$

$(\quad , \quad)$

© Peter Wise, 2015

89

Systems of Equations: Solving by Substitution

1.
$$2x + 3y = 4$$
$$y = \boxed{2x + 4}$$

SOLVE FOR X:

← Substitute (2x + 4) for y in the equation above. Write this value in parentheses.

← Distribute and rewrite the equation

← Combine like terms (with both variables and constants)

$$x = \boxed{}$$ ← Divide by the coefficient of x

SOLVE FOR Y:

← Plug in your value for x into either equation. The second equation is simpler, so let's use that one. Put the value for x in parentheses.

> THE POINT WHERE THE TWO LINES CROSS IS THE X– AND Y–VALUE THAT MAKES BOTH EQUATIONS TRUE!

$$y = \boxed{}$$

Point where the two lines cross: (,)

2. $-4x - 2y = 8$ $y = 3x + 16$

SOLVE FOR X:

← Substitute the value of y on the right into the equation on the left. Write this value in parentheses.

← Distribute and rewrite the equation

← Combine like terms (with both variables and constants)

$$x = \boxed{}$$ ← Divide by the coefficient of x

SOLVE FOR Y:

← Plug in your value for x into either equation. Put the value for x in parentheses.

$$y = \boxed{}$$

Point where the two lines cross: (,)

You can use either equation as long as you get
x = some term(s) or y = some term(s)

1. $x + y = 3$ $4x + 2y = 24$

YOU COULD HAVE ALSO GOTTEN THE 'X' ON BY ITSELF BY SUBTRACTING Y' FROM BOTH SIDES!

This equation is simpler, so start with this one. → $\boxed{x + y = 3}$ $y = \boxed{}$ ← Get the y by itself on the left side by subtracting x from both sides.

YOU WANT TO HAVE 'X' OR 'Y' BY ISELF ON ONE SIDE--SO THAT YOU CAN USE THE OTHER SIDE FOR SUBSTITUTION!

SOLVE FOR X:

← Substitute the terms in the box above for y in the equation.. Write this value in parentheses.

← Distribute and rewrite the equation

← Combine x-terms

← Subtract constants to get them all on the right side of the equation

You could also have used the other equation to get a value for x or y:

$4x + 2y = 24$

← Subtract 4x from both sides

← Divide both sides by 2

$x = \boxed{}$ ← Divide by the coefficient of x

SOLVE FOR Y:

← Plug in your value for x into either equation. The first equation is simpler, so use that one.

$y = \boxed{}$ $\boxed{\text{Point where the two lines cross:}\quad (\quad,\quad)}$

Challenge:

Plug the values for x and y back into both equations and see if both equations are true →

$4x + 2y = 24$ $x + y = 3$
$4(\quad) + 2(\quad) = 24$ $(\quad) + (\quad) = 3$

1.

$$2x - y = 4 \qquad 5x - 5y = -10$$

← Start with this equation

← Divide all three terms by a number to get rid of the coefficient of x.

← Add y to both sides. This will leave x by itself on the left side.

SOLVE FOR Y:

← Substitute the terms in the box above for x in the first equation.. Write this value in parentheses.

← Distribute and rewrite the equation

← Combine y-terms

← Add constants to get them all on the right side of the equation

You could also have used the other equation to get a value for x or y:

$$2x - y = 24$$

← Subtract 2x from both sides

← Multiply both sides by -1

YOU COULD ALSO DIVIDE BOTH SIDES BY -1!

$y = \boxed{}$ ← Divide by the coefficient of x

SOLVE FOR X:

← Plug in your value for x into either equation. Use either equation to solve for x. Use parentheses for the value of y that you are substituting

$x = \boxed{}$

x- and y-values that make both equations true: $(\quad , \quad)$

Systems of Equations: Solving by Substitution

Solve the following systems of equations using the SUBSTITUTION method

1. $3x - 3y = 12$
          $-2x + y = 4$

SOLVE FOR X:

SOLVE FOR Y:

$x =$ ☐

$y =$ ☐

x- and y-values that make both equations true: (,)

SOLVE THIS ONE ON YOUR OWN!

2. $x - 4y = -5$
          $4x - 2y = 8$

x- and y-values that make both equations true: (,)

Slopes Less Than & Greater Than One

This page is information only

A.

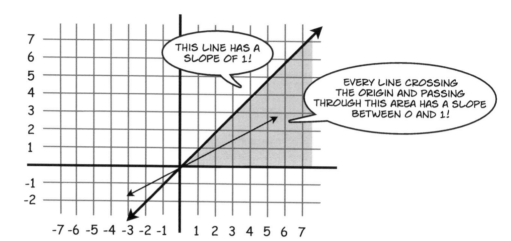

B.

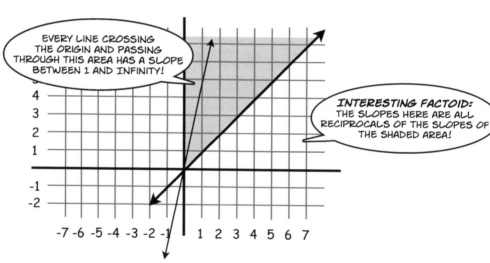

This is a visual proof that there are as many numbers between 0 and 1 as there are between 1 and infinity

C.

Answer Key

Introduction to Graphing in 4 Quadrants

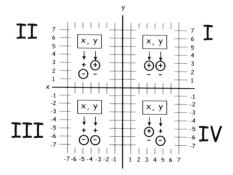

Y-AXIS (VERTICAL)

NOTICE 4 QUADRANTS! THE FIRST ONE STARTS HERE!

THEY GO COUNTER CLOCKWISE!

X-AXIS (HORIZONTAL)

(0,0) IS CALLED THE "ORIGIN"!

Memory Trick!

AN EASY WAY TO REMEMBER THAT THE Y-AXIS IS VERTICAL IS TO THINK OF A LONG TAIL ON THE LETTER "Y"!

oordinates

THINK OF THE DIRECTION OF THE QUADRANTS AS A LETTER "C" FOR "COORDINATES"!

X-AXIS (HORIZONTAL) ↔

If x is positive → ...go to the RIGHT
If x is negative ← ...go to the LEFT

Y-AXIS (VERTICAL) ↕

If y is positive ↑ ...go UP
If y is negative ↓ ...go DOWN

RULE: "WALK AND CLIMB"

IF YOU KICK A BALL INTO A TREE, DO YOU CLIMB IN THE AIR AND WALK? OR WALK TO THE TREE AND CLIMB?

Plot the following points

A. (2,3) WALK to 2
CLIMB to 3

B. (-1,4) WALK (left) to -1
CLIMB to 4

C. (-3,-2) WALK (left) to -3
CLIMB (down) to -2

D. (5,-2) WALK to 5
CLIMB (down) to -2

1

Labeling Points in 4 Quadrants

Plot the following points

1. (3,5) WALK to 3
CLIMB to 5

2. (-2,4) WALK (left) to -2
CLIMB to 4

3. (5,-6) WALK to 5
CLIMB (down) to -6

4. (-4,-3)

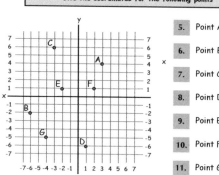

Give the coordinates for the following points

5. Point A (**3** , **4**)

6. Point B (**-6** , **-2**)

7. Point C (**-3** , **6**)

8. Point D (**1** , **-6**)

9. Point E (**-2** , **1**)

10. Point F (**2** , **1**)

11. Point G (**-4** , **-5**)

2

Ordered Pair Signs in Different Quadrants

1. Label the points in the four quadrants. What do you notice about the relationship between the x- and y-values in quadrants that are diagonal to each other? **The signs in diagonal quadrants are opposite for both x and y**

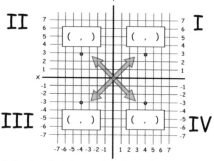

II   I

(,)   (,)

(,)   (,)

III   IV

2. Circle the signs that the x- and y-values will have in each quadrant

II
x, y
↓ ↓
⊖ ⊕

I
x, y
↓ ↓
⊕ ⊕

III
x, y
↓ ↓
⊖ ⊕

IV
x, y
↓ ↓
⊕ ⊖

3

What are Intercepts?

A.

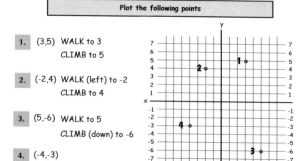

THE Y-INTERCEPT IS -2 (0,-2)

THE X-INTERCEPT IS 3 (3,0)

Intercepts are the points where lines cross through the x- or y-axis

Intercepts always have 0 as the other coordinate. This will be important to remember later on.

SUMMARY:

x-intercept = the point on the on the line where the line crosses the x-axis.

(Notice that the y-value for this will always be 0.)

y-intercept = the point on the on the line where the line crosses the y-axis.

(Notice that the x-value for this will always be 0.)

Plot the points and draw the lines

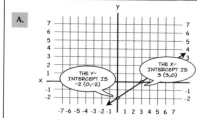

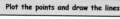

y-intercept = 6   x-intercept = 7   y-intercept = 2   x-intercept = -5

point = (0,6)   point = (7,0)   point (,)   point (,)

4

Introduction to Slope

Most common equation for a line:

A. $y = mx + b$

THINK OF THE M IN "MOUNTAIN SLOPE"

THE "M" IN THIS EQUATION STANDS FOR THE SLOPE!

IT'S THE SLANT OF THE LINE! NOTICE THAT IT MULTIPLES THE x!

Look at two different points in relation to each other. Measure how you get from one to the other.

When you go from one point of a line to another point:

$$\text{Slope is } \frac{\text{Rise}}{\text{Run}} = \frac{\text{Up or down how much?}}{\text{Right or left how much?}}$$

YOU CAN THINK OF "DOWN" AS UP A NEGATIVE AMOUNT!

RIGHT = POS. LEFT = NEG.

Find the slope of the following lines

1.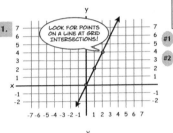

LOOK FOR POINTS ON A LINE AT GRID INTERSECTIONS!

$$\frac{\text{Up or down} \updownarrow}{\text{Left or right} \leftrightarrow}$$

Slope is commonly written as a fraction

#1 Look for points at GRID INTERSECTIONS

#2 Make a FRACTION:

a) Numerator: How much the line goes up/down

  Usually it's easiest to measure points going from left to right; but either way is okay!

b) Denominator: How much the line goes sideways (left/right)

$$\text{Slope} = \frac{\text{Rise}}{\text{Run}} = \frac{2 \updownarrow}{1 \leftrightarrow}$$

2.

#1 Look for places where the line crosses at GRID INTERSECTIONS. Plot points here.

#2 $\dfrac{1 \updownarrow}{3 \leftrightarrow} \dfrac{\text{Rise}}{\text{Run}}$ this is the slope

5

Slope Practice

Find the slope of the following lines

For now, start at the point farthest left

1. Slope = $\dfrac{\text{Up } 2}{\text{Over } 1}$

4. *PLOT AT LEAST TWO POINTS AT GRID INTERSECTIONS!* Slope = $\dfrac{1}{2}$

2. Slope = $\dfrac{\text{Up } 1}{\text{Over } 1}$ or 1

WHAT KIND OF ANGLE IS THIS SLOPE?

5. Slope = $\dfrac{0}{3}$ or 0

WHAT IS THE SLOPE OF ANY HORIZONTAL LINE?

3. $\dfrac{-4}{1}$

WHAT IS DIFFERENT ABOUT THE SLOPE OF THIS LINE?

6. Slope = $\dfrac{4}{3}$

LOOK FOR GRID INTERSECTIONS HERE TOO, AND PLOT TWO POINTS!

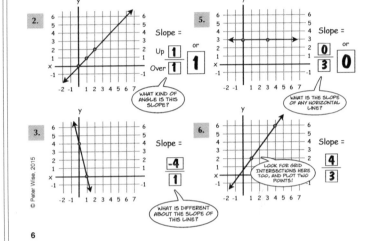
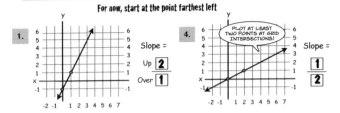

6

Plotting Points and Drawing Lines

Plot points and connect them to draw lines

1. *START HERE!*

Start at the origin and plot 3 points with a slope of $\frac{1}{2}$

$y = \frac{1}{2}x$ slope = $\frac{1}{2}$ (up 1) (right 2)

2.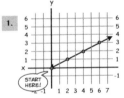

Start at the origin and plot 2 points with a slope of $\frac{2}{3}$

$y = \frac{2}{3}x$ slope = $\frac{2}{3}$ (up 2) (right 3)

3.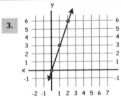

$y = \frac{3}{1}x$ Start at the origin and plot 2 points with a slope of $\frac{3}{1}$

4.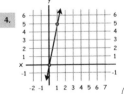

$y = 5x$ Start at the origin and plot 1 more point with a slope of 5

FIRST WRITE THE NUMBER 5 AS A FRACTION!

5.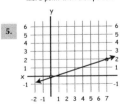

$y = \frac{2}{7}x$ Start at the origin and plot 1 more point with a slope of $\frac{2}{7}$

7

Slope-Intercept Form

Plot points and draw lines

1. y-intercept 2   slope $\frac{1}{4}$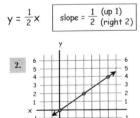

THEN GO UP 1, OVER 4 (THE SLOPE)!

START AT THE Y-INTERCEPT (0,2)!

Start at (0,2)

2. y-intercept -1   slope $\frac{5}{1}$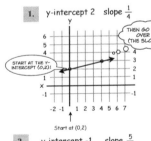

Start at (0,-1)

3. y-intercept 3   slope $\frac{1}{6}$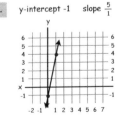

4. y-intercept 1   slope $\frac{4}{5}$

Start at (0,1)

5. y-intercept 0   slope $\frac{5}{3}$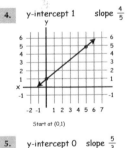

6. y-intercept 3   slope $\frac{2}{7}$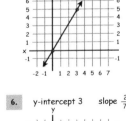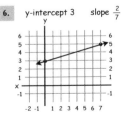

8

Mountain Slope & Balloon

Get acquainted with the various parts of the equation of a line...

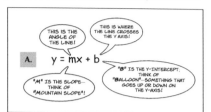

A. $y = mx + b$

- THIS IS THE ANGLE OF THE LINE!
- THIS IS WHERE THE LINE CROSSES THE Y AXIS!
- "M" IS THE SLOPE—THINK OF "MOUNTAIN SLOPE"!
- "B" IS THE Y-INTERCEPT. THINK OF "BALLOON"—SOMETHING THAT GOES UP OR DOWN ON THE Y-AXIS!

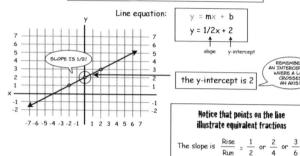

Line equation:
$$y = mx + b$$
$$y = 1/2x + 2$$
slope   y-intercept

SLOPE IS 1/2!

the y-intercept is 2

REMEMBER! AN INTERCEPT IS WHERE A LINE CROSSES AN AXIS!

Notice that points on the line illustrate equivalent fractions

The slope is $\dfrac{Rise}{Run} = \dfrac{1}{2}$ or $\dfrac{2}{4}$ or $\dfrac{3}{6}$

Find the slope and y-intercept of each line:

1. $y = 5x + 3$ slope: **5**   y-intercept: **3**

2. $y = \frac{3}{7}x - 2$ slope: **$\frac{3}{7}$**   y-intercept: **-2**

3. $y = -\frac{4}{9}x - 5$ slope: **$-\frac{4}{9}$**   y-intercept: **-5**

© Peter Wise, 2015

9

Drawing Lines from y = mx + b

How to draw a line with when you have Slope-Intercept form:

1. Put a point on the y-axis for the "b" value.
2. Now look at the "m" value (slope). Start at the point you put for "b" and go UP and OVER for the "m" value.
3. Draw another point. Connect with a line.

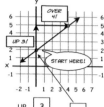

OVER 4!   UP 3!   START HERE!

$$y = \dfrac{UP}{OVER} \boxed{\dfrac{3}{4}} x + \boxed{2}$$

(1) First put a point at (0, 2)

(2) Starting at (0,2) go up 3, over (right) 4 and draw another point.

(3) Connect the points with a line. You're done!

1.

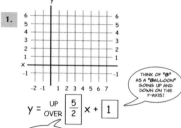

$$y = \dfrac{UP}{OVER} \boxed{\dfrac{5}{2}} x + \boxed{1}$$

THINK OF "B" AS A "BALLOON" GOING UP AND DOWN ON THE Y-AXIS!

THINK OF THIS AS YOUR SECRET MAP DIRECTIONS!

2.

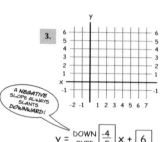

3.

A NEGATIVE SLOPE ALWAYS SLANTS DOWNWARD!

$$y = \dfrac{DOWN}{OVER} \boxed{\dfrac{-4}{5}} x + \boxed{6}$$

$$y = \dfrac{UP}{OVER} \boxed{\dfrac{2}{7}} x - \boxed{1}$$

© Peter Wise, 2015

10

Slope Practice, pt. 2

Start with the first point; use the slope to draw 1-2 more points

Start at (0, -1)   THIS IS THE "Y-INTERCEPT"!

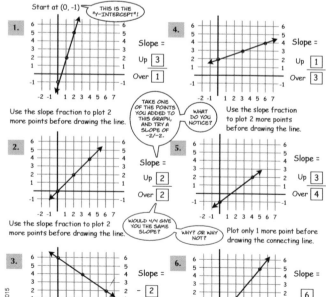

1.   Use the slope fraction to plot 2 more points before drawing the line.

2.   Use the slope fraction to plot 2 more points before drawing the line.

3.   Use the slope fraction to plot 2 more points before drawing the line.

4.   Slope = Up $\boxed{1}$ Over $\boxed{3}$ Use the slope fraction to plot 2 more points before drawing the line.

5.   Slope = Up $\boxed{3}$ Over $\boxed{4}$ Plot only 1 more point before drawing the connecting line.

6.   Plot only 1 more point before drawing the connecting line.

Slope = Up $\boxed{2}$ Over $\boxed{2}$

Slope = $-\dfrac{\boxed{2}}{\boxed{3}}$

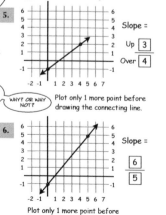

$\dfrac{\boxed{6}}{\boxed{5}}$

TAKE ONE OF THE POINTS YOU ADDED TO THIS GRAPH, AND TRY A SLOPE OF -2/-2.

WHAT DO YOU NOTICE?

WOULD 4/4 GIVE YOU THE SAME SLOPE?

WHY? OR WHY NOT?

© Peter Wise, 2015

11

Negative Slope

Plot points and draw lines

$\dfrac{Negative \; Rise}{Negative \; Run}$ Goes DOWN ↓   Goes to the LEFT ←

1.   Starting at (0,6), plot two points and connect with a line

$slope = \dfrac{-1}{3}$ (down 1) (right 3)

Now, start with your 2nd point and compare it with this slope:

$slope = \dfrac{1}{-3}$ (up 1) (left 3) →

2.   Starting at (1,5), plot two points and connect with a line

$slope = \dfrac{-2}{3}$

Observation about the negative sign:

3.   Starting at (3,4), plot two points and connect with a line

$slope = \dfrac{-1}{2}$

4.   Starting at (-2,5), plot two points and connect with a line

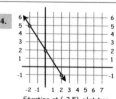

$slope = \dfrac{-3}{2}$

© Peter Wise, 2015

12

Does It Matter Where the Negative Sign Goes?

negative sign can go either on the top or bottom

$$\frac{-3}{1} = -3 \qquad \frac{3}{-1} = -3$$

You can put the negative sign on the numerator ...or the denominator

... same with slope fractions

$$\frac{-2}{3} = \frac{2}{-3}$$

With negative slope fractions you can put the negative sign either on the top or the bottom

YOU'LL GET THE SAME ANSWER EITHER WAY!

Start with the first point; use the slope to draw 1-2 more points

1a.

START HERE- GO DOWN 3 (-3) GO RIGHT 1!

Slope =

Down $\boxed{-3}$

Over $\boxed{1}$

Use the slope to draw one more point

1b.

START HERE.. GO UP 3 GO LEFT 1 (-1)!

Slope =

Up $\boxed{3}$

Left $\boxed{-1}$

Use the slope to draw one more point

Visual proof that both fractions give the same answer

2a.

Slope =

Down $\boxed{-2}$

Over $\boxed{3}$

Start at (-2,6)

Make your first point here; then use the slope to draw 2 more points

2b.

Slope =

Up $\boxed{2}$

Left $\boxed{-3}$

Start at (4,2)

Make your first point here; then use the slope to draw two more points

13

Drawing Rise-Run Arrows

Plot points; draw arrows and lines.

DRAWING SLOPE ARROWS HELPS TO SEE HOW THE LINE IS RELATED TO RISE AND RUN!

A.

ARROWS HAVE BEEN DRAWN TO SHOW A SLOPE OF (-3)/4!

...YOUR TURN! DRAW ARROWS SHOWING A SLOPE OF 3/(-4)!

Start at point C and draw arrows to show that you can get the same result by graphing a slope of $\frac{3}{-4}$.

$$\text{slope} = \frac{3}{-4} \text{ (up 3)} \text{ (left 4)}$$

1. Start at (2,3)

Plot one point with a slope of 2/1, draw rise/run arrows (as in Example A), and connect with a line.

$$\text{slope} = \frac{2}{1}$$

Now start at the same point and draw another point having a slope of (-2)/(-1). Draw arrows.

$$\text{slope} = \frac{-2}{-1}$$

2. Start at (0,0)

Plot three points, draw rise/run arrows, and connect with a line.

$$\text{slope} = \frac{2}{2}$$

Next, start at (0,0) and draw arrows for these two slopes.

$$\text{slope} = \frac{3}{3} \qquad \text{slope} = \frac{5}{5}$$

WHAT DO ALL OF THESE FRACTIONS HAVE IN COMMON?

3. Start at (-2,6)

Plot two points, draw rise/run arrows, and connect with a line.

$$\text{slope} = \frac{-1}{4}$$

14

How Lines are Raised or Lowered

Plot points and connect them to draw lines

Lines on graphs have two parts: (i) Slope (angle), and (2) Height

1.

START HERE!

THIS IS CALLED "Y-INTERCEPT"!

$$y = \boxed{\tfrac{1}{2}} x + 1$$

START UP 1 ON THE Y-AXIS!

#1 Go up 1 on the y-axis and plot your first point (0,1)

#2 Now go up the slope and plot the next point

3.

$$y = \tfrac{3}{4} x + 3$$

#1 Go up this amount on the y-axis and plot your first point

#2 Now go up the slope and plot the next point

2.

$$y = \boxed{\tfrac{3}{1}} x + 2$$

HEIGHT OF THE LINE!

#1 Go up 2 on the y-axis and plot your first point

#2 Now go up the slope and plot the next point

ANGLE (SLOPE) OF THE LINE!

4.

$$y = \tfrac{4}{5} x - 1$$

THIS IS THE "Y-INTERCEPT"!

#1 Look at the value of the y-intercept and go down this amount

#2 Now go up the slope and plot the next point

15

Plug in for 'm' and 'b'

The way of describing a line is called Slope-Intercept Form

$$\boxed{y = mx + b}$$

$$y = (\text{slope})(x) + (\text{y-intercept})$$

"M" IS THE SLOPE

"B" IS THE Y-INTERCEPT!

A.

THINK OF THE "M" AS "MOUNTAIN SLOPE"

"m" = slope

"B" IS THE Y-INTERCEPT!

Y = LINE SLANT + HOW HIGH ON THE Y-AXIS!

$$y = mx + b$$

$$y = \boxed{\tfrac{1}{3}} x + \boxed{4}$$

1.

$$y = \boxed{-\tfrac{1}{4}} x + \boxed{4}$$

Find the slope and y-intercept of the following lines and substitute these values for "m" and "b"

2.

$$y = \boxed{\tfrac{2}{3}} x + \boxed{3}$$

3.

$$y = mx + b \qquad m = \boxed{\tfrac{5}{2}} \quad b = \boxed{-1}$$

16

The Parentheses Trick

- When working with linear equations you will often substitute numbers for variables to plot points
- To substitute a value for a variable—rewrite the equation, PUTTING PARENTHESES in place of that variable (x or y)

 PUT PARENTHESES IN PLACE OF THE VARIABLES!

$y = 3x + 1$

$y = 3(\ ) + 1$

Now pick ANY values for x and solve for y. We started with x = 1

$y = 3(1) + 1$

$y = 4$

x	y
1	

Q. Why do you want to substitute numbers for x?

A. Because this will give you ordered pairs that are on the line

Parentheses provide a helpful "container" for the substituted numbers

Tips for Substituting Numbers

- Plug in any numbers you want for the x-value, and then calculate the y-value
- Often (especially with lines) it's best to plug in x-values that are close to zero: -2, -1, 0, 1, 2, etc.

Rewrite the following equations using parentheses; then plug in two x-values and solve for y

1. $y = 2x + 4$

$y = 2(\ ) + 4$

Rewrite the equation using parentheses for the x-value

x	y
0	4
1	6

3. $y = -3x + 6$

$y = -3(\ ) + 6$

Answers may vary

x	y
0	6
1	3

2. $y = 5x - 5$

$y = 5(\ ) - 5$

x	y
0	-5
1	0

4. $y = -1x - 2$

$y = -1(\ ) - 2$

x	y
0	-2
-1	-1

17

Plot by Making an x/y Table

How to draw a line using an x, y table:

1. Make a blank x,y table
2. Pick easy numbers (like 0, 1, 2) for x. Substitute them for x to figure out the corresponding y-values.
3. Plot the two x, y points and connect them with a line

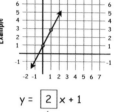
Example

$y = \boxed{2} x + 1$

Same as $\dfrac{2}{1}$

x	y
0	1
1	3

1.

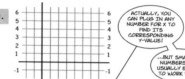 ACTUALLY, YOU CAN PLUG IN ANY NUMBER FOR X TO FIND ITS CORRESPONDING Y-VALUE!

...BUT SMALLER NUMBERS ARE USUALLY EASIER TO WORK WITH!

$y = 3x + 2$

$y = 3(0) + 2$ | 0 | 2 |

$y = 3(1) + 2$ | 1 | 5 |

FIND THE Y VALUE BY PLUGGING IN THE X VALUE!

Answers may vary

2.

$y = -2x + 4$

$y = -2(\ ) + 4$

x	y
1	2
2	0

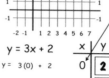 REWRITE THE EQUATION, BUT PUT PARENTHESES UNDER THE X!

3.

$y = \dfrac{1}{5} x + 5$

$y = \dfrac{1}{5}(\ ) + 5$

 USE A VALUE FOR X THAT WILL CANCEL OUT THE DENOMINATOR!

x	y
0	5
5	6

18

Practice with x/y Tables

Plug in the x-values and solve for y

1. $y = x + 2$

Plug in the x values and see what the y values become!

$y = (\ ) + 2$

Plug in the x values here

Solve for y

USING PARENTHESES AS A CONTAINER FOR SUBSTITUTING NUMBERS IS A GOOD PRACTICE, ESPECIALLY IN THE EARLY STAGES OF DOING SUBSTITUTION!

x	y
-2	0
-1	1
1	3
2	4
8	10

2. $y = -x$ same as y = (-1)x

$y = -(\ )$

NOTE! NEGATIVE ONE IS A SIGN SWITCHER!

x	y
-2	2
-1	1
1	-1
2	-2

3. $y = \dfrac{1}{2} x + 1$

$y = \dfrac{1}{2}(\ ) + 1$

With slopes in fraction form, it is easiest to plug in values that are multiples of the denominator—you will always get whole numbers! (This is because the multiples will cancel the denominator.)

x	y
-6	-2
-2	0
0	1
2	2
8	5

4. $y = \dfrac{1}{3} x - 2$

$y = \dfrac{1}{3}(\ ) - 2$

x	y
-6	-4
-3	-3
3	-1
9	1
12	2

5. $y = -\dfrac{3}{5} x + 1$

$y = -\dfrac{3}{5}(\ ) + 1$

x	y
-10	7
-5	4
10	-5
20	-11
30	-17

6. $y = -x + 1$

$y = -(\ ) + 1$

x	y
-2	3
-1	2
1	0
2	-1
6	-5

19

Practice with x/y Tables

Plug in the x-values and solve for y

1. $y = 2x + 1$

$(\ ) = 2(\ ) + 1$

x	y
-2	-3
-1	-1
0	1
1	3
2	5

2. $y = -5x$

$y = -5(\ )$

USE THE PARENTHESES TRICK!

REMEMBER! MULTIPLYING BY A NEGATIVE NUMBER SWITCHES ALL OF THE SIGNS!

x	y
-2	10
-1	5
1	-5
2	-15

3. $y = 3x - 2$

$y = 3(\ ) - 2$

x	y
-2	-8
-1	-5
0	-2
1	1
2	4

4. $y = 5x - 10$

$y = 5(\ ) - 10$

x	y
-4	-30
-2	-20
0	-10
2	0
4	10

5. $y = \dfrac{2}{3} x + 3$

$y = \dfrac{2}{3}(\ ) + 3$

With slopes in fraction form, it is easiest to plug in values that are multiples of the denominator—you will always get whole numbers! (This is because the multiples will cancel the denominator.)

x	y
-6	-1
-3	1
0	3
3	4
6	5

6. $y = \dfrac{4}{5} x - 2$

$y = \dfrac{4}{5}(\ ) - 2$

x	y
-10	-10
-5	-6
0	-2
5	2
10	6

20

Substitution with Fraction Slopes

Example

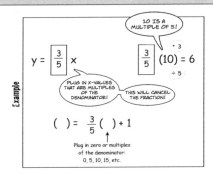

$y = \dfrac{3}{5} x$

10 IS A MULTIPLE OF 5!

$\dfrac{3}{5} (10) = 6$

· 3
÷ 5

PLUG IN X-VALUES THAT ARE MULTIPLES OF THE DENOMINATOR!

THIS WILL CANCEL THE FRACTION!

$( ) = \dfrac{3}{5} ( ) + 1$

Plug in zero or multiples of the denominator: 0, 5, 10, 15, etc.

Rewrite the following equations using parentheses; then plug in two x-values that are MULTIPLES of the DENOMINATOR, and solve for y

1. $y = \dfrac{2}{3} x + 1$

$y = \dfrac{2}{3} ( ) + 1$

x	y
3	3
6	5

Rewrite the equation using parentheses for the x-values; plug in multiples of the denominator (3, 6, 9, 12, etc)

Answers may vary

3. $y = \dfrac{5}{7} x - 2$

$y = \dfrac{5}{7} ( ) - 2$

x	y
7	3
14	8

2. $y = \dfrac{3}{4} x + 2$

$y = \dfrac{3}{4} ( ) + 2$

x	y
4	5
8	8

4. $y = -\dfrac{2}{5} x + 3$

$y = \dfrac{2}{5} ( ) - 2$

x	y
5	0
10	2

21

The y-in-Terms-of-x Trick

Copy the expression that the y term equals and use it instead of the y-value in your table

Example

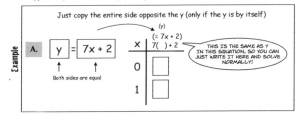

Just copy the entire side opposite the y (only if the y is by itself)

A. $y = 7x + 2$

Both sides are equal

(y) (= 7x + 2)

x	7( ) + 2
0	
1	

THIS IS THE SAME AS y IN THIS EQUATION, SO YOU CAN JUST WRITE IT HERE AND SOLVE NORMALLY!

Make x/y tables, but now write y in terms of x (as in the example above)

y =

1. $y = 4x - 3$

x	4( ) - 3
0	-3
1	1

4. $y = -\dfrac{4}{3} x + 2$

x	$-\dfrac{4}{3}$( ) + 2
3	-2
6	-6

Answers may vary

YOU CAN ELIMINATE FRACTION RESULTS BY PLUGGING IN THE DENOMINATOR OR MULTIPLES OF THE DENOMINATOR!

2. $y = \dfrac{3}{8} x - 2$

x	$\dfrac{3}{8}$( ) - 2
8	1
16	4

5. $y = \dfrac{5}{9} x + 3$

x	$\dfrac{5}{9}$( ) + 3
9	8
18	13

3. $y = -\dfrac{7}{2} x - 1$

x	$-\dfrac{7}{2}$( ) - 1
2	-8
4	-15

6. $y = -\dfrac{8}{5} x + 10$

x	$-\dfrac{8}{5}$( ) + 10
-5	18
10	-6

22

Seeing y-in-Terms-of-x on a Graph

Writing y as a function of x; A.K.A "f of x"

1. $y = 2x$

IN THIS EQUATION, THE Y-VALUES ARE WHAT YOU GET WHEN YOU MULTIPLY X BY 2!

(x,2x)
(2 , 4)
x =   y =
LABEL THE COORDINATES!
(x,2x)
(1 , 2)

2. $y = \dfrac{1}{2} x$

THE HEIGHT (Y-VALUE) WILL BE 1/2 OF THE SIDEWAYS DIRECTION (X-VALUE)!

$(x, \tfrac{1}{2}x)$
(6 , 3)
$(x, \tfrac{1}{2}x)$
(2 , 1)

3. $y = 3x - 2$

(x,3x-2)
(3 , 7)
(x,3x-2)
(0 , -2)

WRITE THE COORDINATES AS ABOVE (Y IN TERMS OF X)

4. $y = 5x$ (x , 5x)

5. $y = \dfrac{4}{5} x + 3$ (x, $\tfrac{4}{5}$x + 3)

6. $y = -\dfrac{1}{4} x - 2$ (x, $-\tfrac{4}{5}$x - 2)

23

Graph and Explain the Difference

What differences do you see between these pairs of linear equations?

1a. $y = 1x + 0$

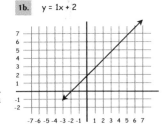

1b. $y = 1x + 2$

Explain the difference:
Both have the same slope, but 1b is 2 units higher

2a. $y = 2x + 2$

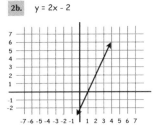

2b. $y = 2x - 2$

Explain the difference:
Both have the same slope, but 1b is 4 units lower

24

Graph the lines and make an observation about the differences in slope

1. $y = 6x + 0$

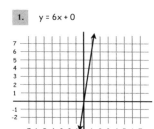

3. $y = \frac{1}{2}x + 0$

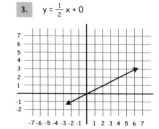

2. $y = 1x + 0$

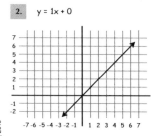

4. $y = \frac{1}{7}x + 0$

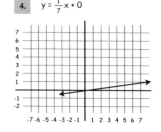

What do you notice in the lines as you go from problem 1 to problem 4?

The larger the slope, the steeper the line. The smaller the slope the flatter it is.

© Peter Wise, 2015

25

Plot the points and draw the lines

1a. $y = 3x + 1$

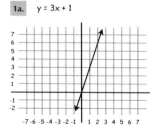

1b. $y = -3x + 1$

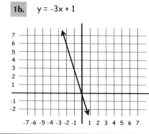

Explain the difference: 1a is a positive slope (going up 3 times as much as it moves to the right. 1b is a negative slope (going DOWN 3 times as much as it moves to the right.

2a. $y = 2x + 1$ **2b.** $y = -\frac{1}{2}x + 1$ **3a.** $y = \frac{2}{3}x + 1$ **3b.** $y = -\frac{3}{2}x + 1$

Plot points and draw lines on the same graph

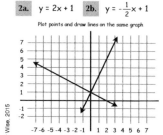

Plot points and draw lines on the same graph

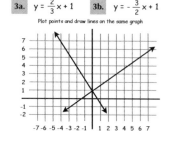

Explain the difference? Be specific!
In both sets the lines are perpendicular. The slopes of the a & b versions are negative reciprocals of each other.

© Peter Wise, 2015

26

Find the slope and the y-intercept of each equation

1. $y = \frac{2}{3}x + 5$ slope = $\boxed{\frac{2}{3}}$ y-intercept = $\boxed{5}$

2. $y = \frac{5}{4}x - 2$ slope = $\boxed{\frac{5}{4}}$ y-intercept = $\boxed{-2}$

3. $y = \frac{6}{1}x - 7$ slope = $\boxed{\frac{6}{1}(6)}$ y-intercept = $\boxed{-7}$

4. $y = \frac{8}{9}x - 1$ slope = $\boxed{\frac{8}{9}}$ y-intercept = $\boxed{-1}$

5. $y = \boxed{\frac{8}{3}}x \boxed{+3}$ slope = $\boxed{\frac{8}{3}}$ y-intercept = $\boxed{3}$

6. $y = \frac{-4}{3}x + 8$ slope = $\boxed{\frac{-4}{3}}$ y-intercept = $\boxed{8}$

7. $y = \boxed{\frac{7}{-5}}x \boxed{-1}$ put + or - here   slope = $\boxed{\frac{7}{-5}}$ y-intercept = $\boxed{-1}$

8. $y = \frac{7}{-11}x$ slope = $\boxed{\frac{7}{-11}}$ y-intercept = $\boxed{0}$

9. $y = \boxed{\frac{1}{5}}x \boxed{+4}$ y-intercept = 4   slope = $\frac{1}{5}$

10. $y = \boxed{-6}x \boxed{-16}$ b = -15   m = -6

© Peter Wise, 2015

27

Determine the slope and the y-intercept from the graphs

SLOPE goes here →   y-intercept goes here →

1. $y = \boxed{\frac{-5}{2}}x \boxed{+} \boxed{6}$
put + or - here

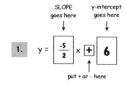

IT IS HELPFUL TO PUT POINTS WHERE THE LINE INTERSECTS BOTH GRID LINES!

2. $y = \boxed{2}x \boxed{+} \boxed{4}$
put + or - here

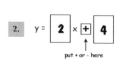

PUT POINTS WHERE THE LINE INTERSECTS THE GRID LINES!

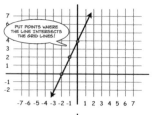

3. $y = \boxed{\frac{-1}{3}}x \boxed{-} \boxed{3}$

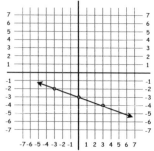

© Peter Wise, 2015

28

To make a line perpendicular to this one, just change the slope:

$$y = \frac{3}{4}x + 2$$

- Use the RECIPROCAL of this slope
- Make it NEGATIVE

$$y = -\frac{4}{3}x + 2$$

THE FRACTION IS FLIPPED AND IT HAS THE OPPOSITE SIGN!

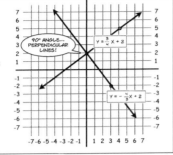

90° ANGLE... PERPENDICULAR LINES!

$y = \frac{3}{4}x + 2$

$y = -\frac{4}{3}x + 2$

Give an equation for a perpendicular line

1. $y = \frac{2}{5}x + 1$

$$y = -\frac{5}{2}x + 1$$

2. $y = 3x + 4$

$$y = -\frac{1}{3}x + 4$$

3. $y = -\frac{7}{3}x + 2$

$$y = \frac{3}{7}x + 2$$

4. $y = -2x + 5$

$$y = \frac{1}{2}x + 5$$

5. $y = -1\frac{1}{3}x - 2$

$$y = \frac{3}{4}x - 2$$

6. $y = 2\frac{3}{4}x + 1$

$$y = -\frac{4}{11}x + 1$$

© Peter Wise, 2015

29

Make the y-value 1y

A. $2y = 6x + 8$ $\dfrac{2y}{2} = \dfrac{6x}{2} + \dfrac{8}{2} \longrightarrow$ $\boxed{y = 3x + 4}$

THE Y-VALUE HAS TO BE ONE—SO GET RID OF THE 2 BY DIVIDING EVERY TERM BY 2!

Example

Rewrite the equations so that the y-value is 1y (or just y)

THIS HAS TO BE 1Y!

1. $5y = 10x + 15$ Divide every term by: $\boxed{5}$

$$\boxed{y = 2x + 3}$$

Rewritten equation

2. $3y = x + 12$ Divide every term by: $\boxed{3}$

$$\boxed{y = \frac{1}{3}x + 4}$$

Rewritten equation

3. $-y = 2x + 10$ Divide every term by: $\boxed{-1}$ (or multiply)

THERE IS AN INVISIBLE NUMBER IN FRONT OF THE Y!

$$\boxed{y = -2x - 10}$$

Rewritten equation

4. $\frac{3}{4}y = 15x + 9$ Multiply every term by: $\boxed{\frac{4}{3}}$

WHAT MAGIC TOOL MAKES FRACTIONS DISAPPEAR BY TURNING THEM TO ONES?

$$\boxed{y = 20x + 12}$$

Rewritten equation

© Peter Wise, 2015

30

$y = mx + b$ is Slope-Intercept Form—you can arrange the equation to match this form

A. $y - 3 = 2x \longrightarrow$ $\boxed{y = 2x + 3}$
   $\quad +3 \quad +3$

GET Y BY ITSELF ON ONE SIDE!

B. $y - 5x = 4 \longrightarrow$ $\boxed{y = 5x + 4}$
   $\quad +5x \ +5x$

Example

Change sides, change signs

Negative 3 on one side of the equal sign = positive 3 on the other side

- 5x on one side of the equal sign = + 5x on the other side

Rewrite the equations so that they are in slope-intercept form (y = mx + b)

1. $y - 10 = 4x$

$$\boxed{y = 4x + 10}$$

2. $y - 3 = 4x + 5$

$$\boxed{y = 4x + 8}$$

3. $y - \frac{2}{5} = 7x + \frac{4}{5}$

$$\boxed{y = 7x + \frac{6}{5}}$$

4. $y + 6 = \frac{3}{7}x + 4$

$$\boxed{y = \frac{3}{7}x - 2}$$

5. $y + 5x = -2x - 6$

$$\boxed{y = -7x - 6}$$

6. $y + 7x = 5x + 1$

$$\boxed{y = -2x + 1}$$

7. $y + 10x - 8 = 5x - 10$

$$\boxed{y = -5x - 2}$$

8. $y - x - 4 = 0$

$$\boxed{y = x + 4}$$

9. $-3x + 7 + y = -11x + 2$

$$\boxed{y = -8x - 5}$$

10. $y + \frac{3}{7}x = \frac{2}{3}x - 4$

$$\boxed{y = \frac{5}{21}x - 4}$$

© Peter Wise, 2015

31

A. Find the slope of the line going through these points: $(8,5)$, $(3,2)$

(THE Y-VALUES ARE ON THE RIGHT SIDE!)

#1 Circle the y-value (second number) in each ordered pair $(8,\textcircled{5})$, $(3,\textcircled{2})$

TO SEE THE SLOPE CLEARLY, YOU NEED THE RIGHT GLASSES!

#2 Put these numbers in the ovals (at the TOP of the fraction) and subtract them

$$\frac{5 - 2}{8 - 3} = \frac{3}{5}$$

The RIGHT numbers wear the GLASSES on the TOP ("head") of the fraction!

#3 Put the x-values on the BOTTOM and subtract them

Example

Find the slope of the following points by SUBTRACTION; use the GLASSES TRICK

THE "GLASSES" ARE CONTAINERS THAT HELP YOU KEEP TRACK OF NEGATIVE NUMBERS AND SUBTRACTION SIGNS!

1. Find the slope of the line going through these points: $(4,6)$, $(1,5)$

HAVING THE SIGNS OF THE NEGATIVE INTEGERS IN HERE HELPS TO KEEP TRACK OF THE SIGNS!

- Circle the y-values
- Put them in the "glasses frames"
- Put the x-values on the bottom boxes

$$\frac{6 - 5}{4 - 1} = \frac{1}{3}$$

- Subtract top and bottom to find the slope (often it will be a fraction)

ACTUALLY, YOU CAN SUBTRACT IN THE REVERSE DIRECTION AND STILL GET THE SAME ANSWER!

$$\frac{5 - 6}{1 - 4} = \frac{\ }{\ } = \frac{\ }{\ }$$

2. Find the slope of the line going through these points: $(3,2)$, $(1,1)$

DRAW GLASSES FOR THE TOP NUMBERS AND BOXES FOR THE BOTTOM NUMBERS!

- Circle the y-values (RIGHT numbers)
- Put these "glasses numbers" on top

$$\frac{2 - 1}{3 - 1} = \frac{1}{2}$$

© Peter Wise, 2015

32

Finding Slope from Two Points

Use the GLASSES TRICK to find the slope of the following points

1. Find the slope of the line going through these points: $(4,2), (1,1)$

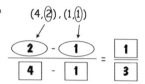

It doesn't matter which point you start with as long as you are consistent

#1 Circle the y-value (second number) in each ordered pair

#2 Put these numbers in the ovals (numerators) and subtract them

#3 Put the x-values on the bottom and subtract them

$(4,\textcircled{2}), (1,\textcircled{1})$

$$\frac{\textcircled{2} - \textcircled{1}}{4 - 1} = \frac{1}{3}$$

2. Find the slope of the line going through these points: $(6,9), (2,6)$

$$\frac{\textcircled{9} - \textcircled{6}}{6 - 2} = \frac{3}{4}$$

CONTINUE USING THE GLASSES TRICK!

3. Find the slope of the line going through these points: $(3,8), (6,3)$

$$\frac{\textcircled{8} - \textcircled{3}}{3 - 6} = \frac{5}{-3}$$

4. Find the slope of the line going through these points: $(-7,-6), (-4,-2)$

$$\frac{\textcircled{-6} - \textcircled{-2}}{-7 - -4} = \frac{4}{3}$$ (+2) / (+4)

THE GLASSES FRAMES HELP YOU KEEP TRACK OF NEGATIVE NUMBERS AND SUBTRACTION SIGNS! neg. signs cancel

© Peter Wise, 2015

33

Finding Slope from Two Points

Use the GLASSES TRICK to find the slope from two points

1. Find the slope of the line going through these points: $(3,5), (2,1)$

Y-VALUES GO ON TOP!

Circle the y-values

$$\frac{\textcircled{5} - \textcircled{1}}{3 - 2} = \frac{4}{1}$$

The x-values go on the bottom

2. Find the slope of the line going through these points: $(4,1), (11,4)$

You can subtract the first point minus second the point or the second point minus the first point, as long as you are consistent

$$\frac{\textcircled{1} - \textcircled{4}}{4 - 11} = \frac{3}{7}$$

neg. signs cancel

3. Find the slope of the line going through these points: $(-1,-14), (-2,10)$

REMEMBER THAT "MINUS A NEGATIVE" = "ADD A POSITIVE"!

$$\frac{\textcircled{-14} - \textcircled{10}}{-1 - -2} = -\frac{24}{1}$$

DRAWING THESE FRAMES HELPS PREVENT LOSING TRACK OF NEGATIVE SIGNS!

4. Find the slope of the line going through these points: $(-7,-4), (-2,7)$

$$\frac{\textcircled{-4} - \textcircled{7}}{-7 - -2} = \frac{11}{5}$$

neg. signs cancel

5. Find the slope of the line going through these points: $(-3,2), (12,-5)$

(+5)

$$\frac{\textcircled{2} - \textcircled{-5}}{-3 - 12} = -\frac{7}{15}$$

© Peter Wise, 2015

34

Finding Slope from Two Points

Find the slope of the following points by SUBTRACTION

1. Find the slope of the line going through these points: $(4,1), (7,5)$

CIRCLE THE Y-VALUES!

$$\frac{\textcircled{1} - \textcircled{5}}{4 - 7} = \frac{4}{3}$$

neg. signs cancel

2. Find the slope of the line going through these points: $(-5,6), (-3,3)$

DRAW IN THE OVAL AND RECTANGLE FRAMES!

$$\frac{\textcircled{6} - \textcircled{3}}{-5 - -3} = -\frac{3}{2}$$

CIRCLE THE Y-VALUES! (+3)

3. Find the slope of the line going through these points: $(-6,-3), (2,4)$

$$\frac{\textcircled{-3} - \textcircled{4}}{-6 - 2} = \frac{7}{8}$$

neg. signs cancel

4. Find the slope of the line going through these points: $(5,-4), (-5,1)$

$$\frac{\textcircled{-4} - \textcircled{1}}{5 - -5} = -\frac{5}{10}$$ (+5)

or -1/2

5. Find the slope of the line going through these points: $(-9,-7), (-4,-1)$

$$\frac{\textcircled{-7} - \textcircled{-1}}{-9 - -4} = \frac{6}{5}$$

neg. signs cancel

6. Find the slope of the line going through these points: $(-7,-2), (4,-5)$

$$\frac{\textcircled{-2} - \textcircled{-5}}{-7 - 4} = -\frac{3}{11}$$

© Peter Wise, 2015

35

Finding Slope from Tables

Find the slope of the following points in the table

1. Find the slope of the line going through these points:

REMEMBER! ONLY Y-VALUES GO HERE ON TOP!

$$\frac{\textcircled{10} - \textcircled{3}}{-4 - (-2)} = -\frac{7}{2}$$

THE TEMPTATION IS TO PUT THE NUMBERS IN THE TOP ROW ON TOP OF THE SLOPE FRACTION, INSIDE THE GLASSES—BUT DON'T DO THIS IF THE TOP ROW HAS THE X-VALUES!

subtract any of these on the BOTTOM

x	-4	-2	0	2
y	10	3	-4	-11

y-values can appear on either the top or bottom rows; be careful!

FIRST CHECK TO SEE WHERE THE Y-VALUES ARE! PUT THEM ON TOP, NO MATTER WHERE THEY APPEAR IN A TABLE!

subtract any of these on the TOP

2. Find the slope of the line going through these points:

$$\frac{\textcircled{14} - \textcircled{4}}{-8 - 4} = \frac{10}{-12} = -\frac{5}{6}$$

simplify

y	14	4	-1	-6
x	-8	4	10	16

3. Find the slope of the line going through these points:

$$\frac{\textcircled{-10} - \textcircled{-7}}{-9 - 1} = \frac{3}{10}$$

x	-9	1	11	21
y	-10	-7	-4	-1

4. Find the slope of the line going through these points:

$$\frac{\textcircled{-22} - \textcircled{-11}}{-16 - -10} = \frac{-11}{-6} = \frac{11}{6}$$

y	-22	-11	0	11
x	-16	-10	-4	2

© Peter Wise, 2015

36

Linear Equations Review

1. Find the slope of the line going through these points: (-6,3),(-4,-2)

CIRCLE THE Y-VALUES!

$$\frac{\boxed{3} - \boxed{-2}}{\boxed{-6} - \boxed{-4}} = -\frac{\boxed{5}}{\boxed{2}}$$

(+ 2) ... (+ 4)

DRAW IN THE OVAL AND RECTANGLE FRAMES!

2. Find the slope of the line going through these points: (-9,-7),(-6,-5)

$$\frac{\boxed{-7} - \boxed{-5}}{\boxed{-9} - \boxed{-6}} = \frac{\boxed{2}}{\boxed{3}}$$

(+ 5) ... (+ 6)

3. Plot three points (one x-value can be zero)

$$y = \frac{1}{5}x + 1$$

$$(\ ) = \frac{1}{5}(\ ) + 1$$

CHOOSE X-VALUES THAT ARE MULTIPLES OF 5 TO ELIMINATE FRACTIONS!

x	y
0	1
5	2
10	3

4. Negative slopes always go:

Upward (Downward) *(Circle one)*

...as you move from left to right on a graph

5. Graph the line: y = -2/3x + 4

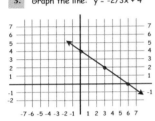

6. Give the equation of the line

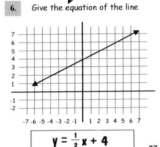

$$y = \frac{1}{2}x + 4$$

© Peter Wise, 2015

37

Equations from Slope and One Point

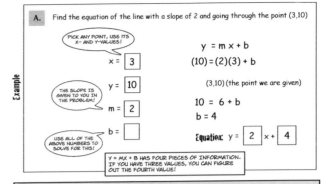

A. Find the equation of the line with a slope of 2 and going through the point (3,10)

PICK ANY POINT, USE ITS X- AND Y-VALUES!

x = 3

THE SLOPE IS GIVEN TO YOU IN THE PROBLEM!

y = 10

m = 2

USE ALL OF THE ABOVE NUMBERS TO SOLVE FOR THIS!

b = ☐

$$y = mx + b$$
$$(10) = (2)(3) + b$$

(3,10) (the point we are given)

$$10 = 6 + b$$
$$b = 4$$

Equation: y = $\boxed{2}$ x + $\boxed{4}$

Y = MX + B HAS FOUR PIECES OF INFORMATION. IF YOU HAVE THREE VALUES, YOU CAN FIGURE OUT THE FOURTH VALUE!

Find the equations of the following lines

1. Find the equation of the line with a slope of 1/2 and going through the point (4,5)

x = $\boxed{4}$ m = $\boxed{\frac{1}{2}}$

PLUG IN THE VALUES!

y = $\boxed{5}$ b = $\boxed{3}$

IF YOU HAVE THREE VALUES YOU CAN FIGURE OUT THE FOURTH!

$$y = mx + b$$
$$(\ ) = (\ )(\ ) + b$$

LINEAR EQUATIONS NEED TO HAVE JUST "X" AND "Y" AS VARIABLES!

Equation: y = $\boxed{\frac{1}{2}}$ x + $\boxed{3}$

2. Find the equation of the line with a slope of 3 and going through the point (1,7)

x = $\boxed{1}$ m = $\boxed{3}$

y = $\boxed{7}$ b = $\boxed{4}$

$$(\ ) = (\ )(\ ) + b$$

Equation: y = $\boxed{3}$ x + $\boxed{4}$

3. Find the equation of the line with a slope of 2 and going through the point (-5,8)

x = $\boxed{-5}$ m = $\boxed{2}$

y = $\boxed{8}$ b = $\boxed{18}$

$$(\ ) = (\ )(\ ) + b$$

Equation: y = $\boxed{2}$ x + $\boxed{18}$

38

Parallel Lines Have the Same Slope

A. Find the equation of the line going through the point (3,5) and parallel to the line $y = \frac{2}{3}x + 2$

x = $\boxed{3}$

PARALLEL LINES HAVE THE SAME SLOPE!

y = $\boxed{5}$

m = $\boxed{\frac{2}{3}}$

THE NEW LINE HAS THE POINT (3,5) AND HAS THE SAME SLOPE AS ANY LINE PARALLEL TO IT!

b = $\boxed{?}$

This is the variable you have to figure out. If you know the values of three variables you can figure out the value of the fourth one!

$$y = mx + b$$
$$5 = \frac{2}{3}(3) + b$$
$$5 = 2 + b$$
$$b = 3$$

MAKE YOUR EQUATION TO SOLVE FOR THE B-TERM!

NOW WRITE THE EQUATION WITH ONLY X AND Y AS VARIABLES!

$$y = \frac{2}{3}x + 3$$

Write equations for the following lines (solve for a new "b" value)

1. Find the equation of the line going through the point (12,7) and parallel to the line $y = \frac{1}{3}x + 8$

x = $\boxed{6}$ m = $\boxed{\frac{1}{3}}$

$$7 = \frac{1}{3}12 + b$$

y = $\boxed{7}$ b = $\boxed{3}$

$$3 = b$$

$$y = \boxed{\frac{1}{3}}x + \boxed{3}$$

2. Find the equation of the line going through the point (-8,10) and parallel to the line $y = -\frac{3}{4}x + 4$

x = $\boxed{-8}$ m = $\boxed{-\frac{3}{4}}$

$$10 = -8\left(-\frac{3}{4}\right) + b$$

y = $\boxed{10}$ b = $\boxed{4}$

$$10 = 6 + b$$
$$10 = 4$$

$$y = -\frac{3}{4}x + 4$$

© Peter Wise, 2015

39

Parallel Lines Have the Same Slope

Write equations for the following lines

1. Find the equation of the line going through the point (-6,-1) and parallel to the line $y = -\frac{2}{3}x - 3$

x = $\boxed{-6}$ m = $\boxed{-\frac{2}{3}}$

y = $\boxed{-1}$ b = $\boxed{-5}$

This is the variable you have to figure out, given the slope and the points (-6,-1)

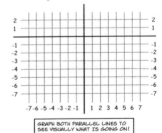

Equation:

$$y = -\frac{2}{3}x - 5$$

GRAPH BOTH PARALLEL LINES TO SEE VISUALLY WHAT IS GOING ON!

2. Find the equation of the line going through the point (-10,10) and parallel to the line $y = -\frac{3}{5}x - 5$

x = $\boxed{-10}$ m = $\boxed{-\frac{3}{5}}$

y = $\boxed{10}$ b = $\boxed{4}$

Equation:

$$y = -\frac{3}{5}x + 4$$

3. Find the equation of the line going through the point (-8,-7) and parallel to the line $y = -\frac{3}{2}x - 2$

Make a list of the four variables as above

x = $\boxed{-8}$ m = $\boxed{-\frac{3}{2}}$

y = $\boxed{-7}$ b = $\boxed{-19}$

NOTE! ANSWERS AREN'T ALWAYS WHOLE NUMBERS!

Equation:

$$y = -\frac{3}{2}x - 19$$

© Peter Wise, 2015

40

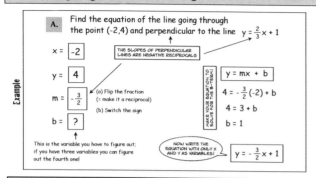

Example

A. Find the equation of the line going through the point (-2,4) and perpendicular to the line $y = \frac{2}{3}x + 1$

$x = \boxed{-2}$

$y = \boxed{4}$

THE SLOPES OF PERPENDICULAR LINES ARE NEGATIVE RECIPROCALS

$m = \boxed{-\frac{3}{2}}$

(a) Flip the fraction (= make it a reciprocal)

(b) Switch the sign

$b = \boxed{?}$

This is the variable you have to figure out; if you have three variables you can figure out the fourth one!

MAKE YOUR EQUATION TO SOLVE FOR THE "B-TERM"

$y = mx + b$

$4 = -\frac{3}{2}(-2) + b$

$4 = 3 + b$

$b = 1$

NOW WRITE THE EQUATION WITH ONLY X AND Y AS VARIABLES!

$y = -\frac{3}{2}x + 1$

Write equations for the following lines

1. Find the equation of the line going through the point (5,6) and perpendicular to the line $y = -\frac{5}{4}x + 1$

$x = \boxed{5}$ $m = \boxed{\frac{4}{5}}$ (a) Flip the slope fraction (= make it a reciprocal)

$y = \boxed{6}$ $b = \boxed{2}$ (b) Switch the sign

Equation:

$y = \frac{4}{5}x + 2$

2. Find the equation of the line going through the point (-3,12) and perpendicular to the line $y = \frac{3}{7}x - 6$

REMEMBER TO GIVE THE SLOPE OF THE NEW (PERPENDICULAR) LINE!

$x = \boxed{-3}$ $m = \boxed{-\frac{7}{3}}$

$y = \boxed{12}$ $b = \boxed{5}$

Equation:

$y = -\frac{7}{3}x + 5$

41

© Peter Wise 2015

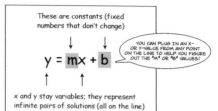

These are constants (fixed numbers that don't change)

$$y = mx + b$$

YOU CAN PLUG IN AN X- OR Y-VALUE FROM ANY POINT ON THE LINE TO HELP YOU FIGURE OUT THE "M" OR "B" VALUES!

x and y stay variables; they represent infinite pairs of solutions (all on the line)

THE X- AND Y-VALUES ON THE LINE CAN EVEN BE FRACTIONS BETWEEN THE WHOLE NUMBERS ON THE LINE—THEY WILL STILL MAKE THE EQUATION TRUE!

These numbers are fixed for each line. They will never change.

$$y = 2x + 1$$

Every point on the line can be substituted for x and y; the equation with the values from these points will always be true

This equation is true if x = 0 and y = 1, or if x = 1 and y = 3, etc. (infinite pairs of solutions)

A. $y = 3x - 1$

Three points on this line are shown. Plug them into the equation to show that POINTS ON A LINE make LINEAR EQUATIONS TRUE

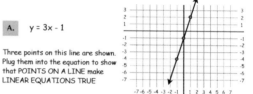

Plug in the values for the point (-1,-4)

Plug in the values for the point (0,-1)

Plug in the values for the point (1,2)

$(-4) = 3(-1) - 1$ $(-1) = 3(0) - 1$ $(2) = 3(1) - 1$

Did the slope change?   Did the y-intercept change?

42

© Peter Wise 2015

---

1. $y = \frac{1}{2}x + 1$

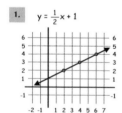

Plug in the points from line into the equation and show that they are valid solutions:

Point (2,2) $(2) = \frac{1}{2}(2) + 1$

Point (4,3) $(3) = \frac{1}{2}(4) + 1$

Point (6,4) $(4) = \frac{1}{2}(6) + 1$

2. $y = 2x - 3$

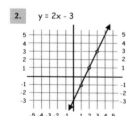

Plug in the points from line into the equation and show that they are valid solutions:

Point (1,-1) $(-1) = (2)(1) - 3$

Point (2,1) $(1) = (2)(2) - 3$

Point (3,3) $(3) = (2)(3) - 3$

3. $y = \frac{2}{3}x - 1$

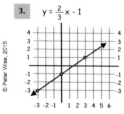

Plug in the points from line into the equation and show that they are valid solutions:

Point (0,-1) $(-1) = (\frac{2}{3})(0) - 1$

Point (3,1) $(1) = (\frac{2}{3})(3) - 1$

43

© Peter Wise 2015

A. Find the y-intercept of a line that has a slope of 2 and the point (3,8)

- The slope is 2
- The point (3,8) is on the line

You know 3 out of 4 things. Your task is to find the value of "b"

$x = \boxed{3}$

$y = \boxed{8}$

$m = \boxed{2}$

THIS IS WHAT WE'RE LOOKING FOR!

$b = \boxed{}$

If you have any three of these values you can figure out the remaining one

$y = mx + ⓑ$

$8 = 2(3) + b$

$b = \boxed{2}$

---

1. A line has a slope of 3 and one point on the line is (4,13). What is the y-intercept?

$x = \boxed{4}$ $m = \boxed{3}$

$y = \boxed{13}$ Plug in these values here

$y = mx + b$

$(13) = (3)(4) + (b)$

$b = \boxed{1}$

2. A line has a slope of -2 and one point on the line is (3,2). What is the y-intercept?

$x = \boxed{3}$ $m = \boxed{-2}$

$y = \boxed{2}$

$(2) = (-2)(3) + (b)$

$b = \boxed{8}$

3. A line has a slope of 4 and one point on the line is (2,5). What is the y-intercept?

$x = \boxed{2}$ $m = \boxed{4}$

$y = \boxed{5}$

$b = \boxed{-3}$

44

© Peter Wise 2015

How to Find Any x- or y-Intercept

Look closely at the x- and y-axis. Find the zero next to the x-axis (representing y—because to be exactly on the x-axis the y-value has to be zero—not up or down any amount!).

Find the y-axis. Look at the zero at the bottom—otherwise the points would be off to the left or the right of the y-axis.

To find a y-intercept, make $x = 0$ and solve for y
To find an x-intercept, make $y = 0$ and solve for x

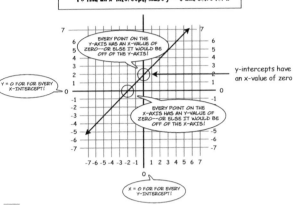

EVERY POINT ON THE Y-AXIS HAS AN X-VALUE OF ZERO—OR ELSE IT WOULD BE OFF OF THE Y-AXIS!

Y = 0 FOR FOR EVERY X-INTERCEPT!

y-intercepts have an x-value of zero

EVERY POINT ON THE X-AXIS HAS AN Y-VALUE OF ZERO—OR ELSE IT WOULD BE OFF OF THE X-AXIS!

X = 0 FOR FOR EVERY Y-INTERCEPT!

1. $y = 2x - 4$

MAKE Y ZERO TO FIND THE X-INTERCEPT!

MAKE X ZERO TO FIND THE Y-INTERCEPT!

$(0) = 2x - 4$ x-intercept = **2** $y = 2(0) - 4$ y-intercept = **-4**

© Peter Wise, 2015

45

Finding Intercepts

To find a y-intercept, make $x = 0$ and solve for y
To find a x-intercept, make $y = 0$ and solve for x

TO FIND ANY INTERCEPT, MAKE THE OPPOSITE COORDINATE ZERO!

RECOMMENDATION: CHECK YOUR ANSWERS ON A GRAPH AFTERWARDS!

1. $y = 4x - 8$

   x-intercept = $(0) = 4x - 8$ **2** x-intercept (**2**, 0) point

TO FIND THE X-INTERCEPT, PLUG IN ZERO FOR Y AND SOLVE!

   y-intercept = $y = 4(0) - 8$ **-8** y-intercept (0, **-8**) point

TO FIND THE Y-INTERCEPT, PLUG IN ZERO FOR X AND SOLVE!

Any intercept has to have one value be zero

Make the opposite coordinate zero and solve!

2. $y = \frac{1}{3}x + 3$

→ x-intercept = $(0) = \frac{1}{3}x + 3$ **-9** x-intercept (-9,0) point

→ y-intercept = $y = \frac{1}{3}(0) + 3$ **3** y-intercept (0,3) point

3. $y = \frac{3}{4}x - 6$

RECOMMENDATION: USE PARENTHESES FOR THE VARIABLE THAT YOU WILL SUBSTITUTE ZERO INTO!

   x-intercept = $(0) = \frac{3}{4}x - 6$ **8** x-intercept (8,0) point

   y-intercept = $y = \frac{3}{4}(0) - 6$ **-6** y-intercept (0,-6) point

4. $y = -\frac{1}{2}x - \frac{7}{2}$

   x-intercept = $(0) = -\frac{1}{2}x - \frac{7}{2}$ **-7** x-intercept (-7,0) point

   y-intercept = $y = -\frac{1}{2}(0) - \frac{7}{2}$ **$-\frac{7}{2}$** y-intercept $\left(0, -\frac{7}{2}\right)$ point

© Peter Wise, 2015

46

Solving for Intercepts

Solve for intercepts by plugging in 0 for x or y

A. $y = 2x + 4$

TO FIND THE X-INTERCEPT, PLUG IN ZERO FOR Y!

TO FIND THE Y-INTERCEPT, PLUG IN ZERO FOR X!

WHEN Y = 0, YOU FIND THE X-INTERCEPT!

$x = (-2)$!   = THE POINT (-2,0)!

To find the x-intercept → $0 = 2x + 4$

To find the y-intercept → $y = 2(0) + 4$ $Y = 4$!   = THE POINT (0,4)

ZERO FOR ONE COORDINATE FINDS THE INTERCEPT FOR THE OTHER!

1. $y = 6x + 12$ x-intercept **-2**   y-intercept **12**

2. $y = 3x + 18$ x-intercept **-6**   y-intercept **18**

3. $y = x + 5$ x-intercept **-5**   y-intercept **5**

4. $y = -4x + 4$ x-intercept **1**   y-intercept **4**

5. $y = \frac{11}{3}x - 11$ x-intercept **3**   y-intercept **-11**

© Peter Wise, 2015

47

Lines When x or y are Missing

Missing an x-value = parallel to x-axis
Missing a y-value = parallel to y-axis

A. Graph the line of the equation $x = 5$

Example

RULE: If you are missing the y-value, the line will be parallel to the y-axis.

THE X-VALUES ARE ONLY AND ALWAYS 5!

x	y
5		**Any numbers**
5	
5	

PUT ANY NUMBERS YOU WANT HERE!

(5,6)   (5,1)   (5,0)

$x = 5$ Don't see a y anywhere? The line will be parallel to the y-axis!

B. Graph the line of the equation $y = 2$

Example

RULE: If you are missing the x-value, the line will be parallel to the x-axis

Put in any numbers you want here!

x	y
	2
	2
	2

Any numbers

(0,2)   (3,2)   (6,2)

$y = 2$ Don't see an x anywhere? The line will be parallel to the x-axis!

1. Plug in values for the equation $x = 6$ and then graph the line

x	y
6	
6	
6	

Any numbers

© Peter Wise, 2015

48

Practice with Missing x- or y-Values

They really aren't missing—you just don't always need to write them

A. Graph the line of the equation y = 3

THE LINE *HEIGHT* (Y-VALUE) IS ALWAYS 3!

Note: If the x-value is missing (really zero) the line is parallel to the x-axis!

EVERY POINT ON THE LINE HAS AN X-COORDINATE OF 4!

Example

THE LINE IS PARALLEL TO THE X-AXIS! THE HEIGHT IS ALWAYS 3!

(0,3)   (5,3)   (2,3)

B. Graph the line of the equation x = 4

(4,5)   (4,3)   (4,1)

YOU DON'T NEED THE Y-COORDINATE BECAUSE X IS ALWAYS 4!

Note: If the y-value is missing (really zero) the line is parallel to the y-axis!

1.   Graph x = 2

Plot several points in which x = 2 (y can be anything else)

REMEMBER! THE Y-TERM IS MISSING; THE LINE WILL BE PARALLEL TO THE Y-AXIS!

3.

THE LINE IS PARALLEL TO THE X-AXIS, SO X-VALUE IS MISSING

Give the equation of the line   y = 1

2.   Graph y = 6

Plot several points in which y = 6 (x can be anything else)

4.

Give the equation of the line   x = 5

49

Finding Slope from x/y Tables (Vertical)

1. Find the slope from the given x/y table

x	y
-4	1
-2	2
0	3
2	4
4	5

PICK ANY TWO RANDOM POINTS AND SUBTRACT Y'S; THEN SUBTRACT X'S!

REMEMBER! IT DOESN'T MATTER WHICH X OR Y YOU START WITH AS LONG AS YOU ARE CONSISTENT!

$$\frac{y \text{ from one point minus the other } y}{x \text{ from the same point minus the other } x} \quad \frac{Y_2 - Y_1}{X_2 - X_1}$$

$Y_2 - Y_1$

$\boxed{3} - \boxed{4} = \boxed{-1}$

$\boxed{0} - \boxed{2} = \boxed{-2}$

$X_2 - X_1$

numbers will vary

$$SLOPE = \frac{1}{2}$$

(negative signs cancel)

2. Find the slope

x	y
-6	2
-3	4
0	6
3	8
6	10

$Y_2 - Y_1$

$\boxed{2} - \boxed{4} = \boxed{-2}$

$\boxed{-6} - \boxed{-3} = \boxed{-3}$

$X_2 - X_1$

$$SLOPE = \frac{2}{3}$$

NOW TRY OTHER POINTS!

$Y_2 - Y_1$

$\boxed{8} - \boxed{6} = \boxed{2}$

$\boxed{3} - \boxed{0} = \boxed{3}$

$X_2 - X_1$

$$SLOPE = \frac{2}{3}$$

3.

x	y
-20	2
-15	4
-10	6
-5	8

$\boxed{8} - \boxed{6} = \boxed{2}$

$\boxed{-5} - \boxed{-10} = \boxed{5}$

$$SLOPE = \frac{2}{5}$$

4.

x	y
-20	2
-16	4
-12	6

$\boxed{4} - \boxed{6} = \boxed{-2}$

$\boxed{-16} - \boxed{-12} = \boxed{-4}$

$$SLOPE = \frac{1}{2}$$

50

Linear Equations Review

REMEMBER *THE GLASSES TRICK!*

DRAW IN THE OVAL AND RECTANGLE FRAMES!

1. Find the slope of the line going through these points: (-10,-4), (-3,-1)

$$\frac{\boxed{-4} - \boxed{-1}}{\boxed{-10} - \boxed{-3}} = \frac{3}{7}$$

neg. signs cancel

2. Plot three points (one x-value can be zero)

$$y = \frac{2}{3}x + 1$$

$$(\ ) = \frac{2}{3}(\ ) + 1$$

CHOOSE X-VALUES THAT ARE MULTIPLES OF 3 TO ELIMINATE FRACTIONS!

x	y
0	1
3	2
6	3

3. Slope is

Change in   **y**

Change in   **x**

4. Find the equation of the line with a slope of $\frac{3}{5}$ and going through the point (-5,1)

Remember to make a list of x, y, m, b

$$y = \frac{3}{5}x + 4$$

5. Find the y-intercept (the b-value)

y = mx + b   b = **-4**

$$(2) = (\tfrac{2}{7})(21) + (-4)$$

- The slope is $\frac{2}{7}$
- The point (21,2) is on the line

6. Find the x-intercept in the line $y = \frac{3}{4}x - 6$ x-intercept = **8**

$$0 = \frac{3}{4}x - 6 \quad 6 = \frac{3}{4}x \quad \frac{4}{4} \cdot 6 = \frac{3}{4}x \cdot \frac{4}{3}$$

7. In the formula y = mx + b, which values are fixed numbers?   **m, b**

Which values are infinite pairs of solutions?   **x, y**

51

---

f(x) as Another Name for the y-Value

f(x) is read "f of x"

A. Linear equations like y = 2x + 1 are commonly written with f(x) instead of y in the equation.

f(x) = 2x + 1

SAME THINGS AS Y!

y = 2x + 1

same thing!

f(x) = 2x + 1

f(x) is called a **function**; this will be covered in later pages!

THINK OF Y SHOWING SPECIAL *F(X)*! (SPECIAL EFFECTS)

...although you can also write this as g(x) or h(x), etc.

So don't have a heart attack if you see f(x). Just think of it as another way of writing y!

1. "f of x" $f(x) = \frac{1}{3}x + 1$

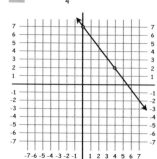

3. $f(x) = -\frac{5}{4}x + 4$

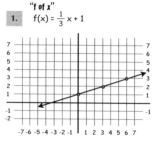

2. $f(x) = \frac{2}{3}x - 2$

Input multiples of 3 to cancel the fraction denominator!

x	3	6	9	12
y	0	2	4	6

52

© Peter Wise, 2015

Graph the following lines

f(x) HERE IS THE SAME AS Y!

1. $f(x) = 3x - 7$

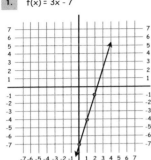

3. $f(x) = \frac{3}{4}x + 2$

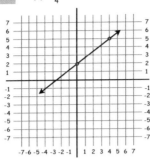

2. $f(x) = 2x - 5$

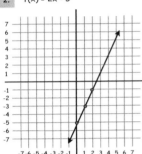

4. $f(x) = -\frac{3}{4}x + 6$

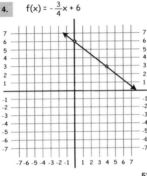

© Peter Wise, 2015

53

---

1. Which part(s) of $y = mx + b$ have fixed numerical values?

____ **m and b** ____

2. Which part(s) of $y = mx + b$ represent infinite pairs of solutions?

____ **x and y** ____

3. If you see the equation: $y = 3$ how would you write this using $y = mx + b$?

____ **$y = (0)x + 3$** ____

What is the slope? ____ **0** ____

Graph the line:

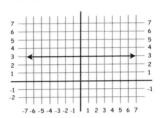

4. What are the two ways of plotting points with the formula $y = mx + b$?

1) Make an x/y table; input any values for x and solve for y

2) Start at the y-intercept; go up the slope

5. $f(x)$ is the same as: **y, the value you get after inputting x**

$f(x)$ is read: **"f of x"**

© Peter Wise, 2015

54

Functions: Only ONE (unique) y-value for any x-value you input

Basics of an x/y Table

x	y
INPUT	OUTPUT
"domain"	"range"
independent variable	dependent variable
-4	1
-2	2
0	3

input
↓
function
↓
output

THIS COLUMN CONSISTS OF INDEPENDENT VARIABLES (= YOU JUST START WITH SOME NUMBERS INDEPENDENTLY!)

THIS COLUMN'S NUMBERS DEPEND ON WHAT YOU PLUGGED IN FOR THE X-VALUE!

DEFINITION OF A FUNCTION: A relationship between a set of x- and y-values in which for every x-value, there is only ONE unique y-value

Just look for duplicate x-values. If the same x-values always give the same y-values, it is a function. (Ignore duplicate y-values.)

This is a Function:

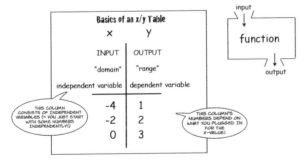

#1 Look for duplicate x-values

(If there are no duplicate x-values, it is a function)

x	y
2	4
5	25
2	4
3	36
6	36

#2 If the duplicate x-values produce the same y-values, then it IS a function

If there are duplicate y-values, just ignore them

© Peter Wise, 2015

55

Functions: Only ONE (unique) y-value for any x-value you input

Visual Characteristic: No two points on the graph of a function can line up vertically

Okay for a function:

DIFFERENT → SAME
x   y

= two ways of getting the same (predictable) results

Not okay for a function:

SAME → DIFFERENT
x   y

= unpredictable results

How to spot non-functions with tables:

Look for SD! ("same → different")

x	y
4	10
2	6
4	12

You can think of SD as "Same Dysfunction"!

LOOK TO SEE IF THE SAME X-VALUE ANYWHERE GIVES TWO DIFFERENT RESULTS FOR Y!

LOOK FOR DUPLICATE X-VALUES! IF THE SAME X-VALUES PRODUCE TWO DIFFERENT Y-VALUES, THEN THIS IS NOT A FUNCTION!

How to spot non-functions visually:

VERTICAL LINE TEST!

MANY SHAPES (LIKE CIRCLES) ARE VALID, BUT JUST AREN'T FUNCTIONS!

IF TWO POINTS LINE UP VERTICALLY, YOU KNOW THIS IS NOT A FUNCTION!

YOU INPUT 5 (X-VALUE), BUT GOT TWO RESULTS (2 AND 6)!

How to spot non-functions with lists of points:

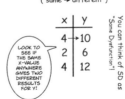

different results

$(2,3)(4,5)(6,2)(7,8)(4,9)$

same input

YOU HAVE TO HAVE DUPLICATE X-VALUES FOR A NON-FUNCTION!

x	1	2	3	2
y	4	7	10	5

#1 Look for two of the same x-values

#2 Check to see if the y-values are different. If they are, then it isn't a function! (unpredictable results)

© Peter Wise, 2015

56

Determine if the following values are functions

#1 Look for duplicate x-values

(If there are no duplicate x-values, it is a function)

#2 If the duplicate x-values produce the SAME y-values, then it IS a function

If the duplicate x-values produce the DIFFERENT y-values, then it is NOT a function

If there are duplicate y-values, just ignore them

1.

x	y
3	5
8	20
2	7
7	8
8	20

(function)   not a function

circle one

2.

x	y
9	4
12	7
2	4
15	8
20	9

(function)   not a function

circle one

3.

x	8	14	7	8	15
y	5	8	4	3	9

function   (not a function)

4.

y	8	13	8	12	17
x	6	9	12	8	11

(function)   not a function

NOTICE THAT THE X- AND Y-VALUES ARE SWITCHED ON THIS PROBLEM!

5. (4,7)(5,8)(5,8)(10,14)(4,9)(6,11)

function   (not a function)

6.

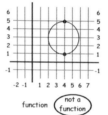

function   (not a function)

Tell if the following sets of points are functions

1.

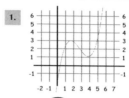

(Yes)   No

(circle one)

Careful! Duplicate y-values for different x-values are okay.

...JUST LIKE GETTING THE SAME ANSWER (Y-VALUE) TWO DIFFERENT WAYS!

5. (3,5)(7,10)(3,6)(8,11)

Yes   (No)

2.

x	5	7	10	7
y	9	11	14	14

Yes   (No)

6.

x	y
-5	25
-2	4
1	1
-2	4

(Yes)   No

3. (2,3)(4,6)(5,8)(7,3)

(Yes)   No

4.

x	y
-7	42
-5	21
-3	4
-1	-4
-3	-2

Yes   (No)

7.

x	-3	1	3	4
y	9	1	9	16

(Yes)   No

For all these use the formula: $y = mx + b$

If you are just missing one coordinate, plug in everything else you have and solve for the missing x- or y-value

TIP: There are four possible pieces of information (x, y, m, b)

If you are missing 1 value the other 3 will enable you to figure out the 4th!

1. Find the value of the y-coordinate if
- the slope is 2
- the y-intercept is 0
- one point is (4,y)

$y = (2)(4) + (0)$

DO YOU REMEMBER WHICH VALUE THE Y-INTERCEPT IS?

x = **4**   m = **2**   b = **0**

$y = $ **8**

2. Find the value of the x-coordinate if
- the slope is 3
- the y-intercept is 2
- one point is (x,8)

$8 = (3)(x) + (2)$

y = **8**   m = **3**   b = **2**

$x = $ **2**

3. Find the value of the y-coordinate if
- the slope is 2
- the y-intercept is -1
- one point is (4,y)

$y = (2)(4) + (-1)$

x = **4**   m = **2**   b = **1**

$y = $ **7**

4. Find the value of the x-coordinate if
- the slope is $\frac{2}{3}$
- the y-intercept is 4
- one point is (x,6)

TO GET RID OF THE 2/3, MULTIPLY BOTH SIDES BY THE DENOMINATOR!

$6 = (\frac{2}{3})(x) + (4)$

$2 = (\frac{2}{3})(x)$

y = **6**   m = $\frac{2}{3}$ b = **4**

multiply both sides by $\frac{3}{2}$

$x = $ **3**

Write equations from the two ordered pairs given

1. Find the equation of the line going through the points (1,5) and (2,9)

PICK ANY POINT YOU ARE GIVEN AND USE THE SAME X- AND Y-VALUES FOR THIS!

HOW MANY PIECES OF INFORMATION DO YOU HAVE FOR THE Y = MX + B FORMULA?

x = **1**

y = **5**

m = **4** or $\frac{4}{1}$

b = **1**

SUBTRACT Y-VALUES (NUMERATOR) AND SUBTRACT THE X-VALUES (DENOMINATOR)!

ONCE YOU HAVE ALL THE OTHER VALUES, USE BASIC ALGEBRA TO SOLVE FOR "B"!

YOU HAVE AN X-VALUE, A Y-VALUE, AND SLOPE (AFTER YOU SUBTRACT THE Y-COORDINATES AND THE X-COORDINATES!)

YOUR EQUATION WILL NOW NEED TO HAVE "X" AND "Y" AS VARIABLES AND "M" AND "B" AS NUMBERS!

solution:

$y = $ **4** $x + $ **1**

2. Find the equation of the line going through the points (2,3) and (4,4)

Pick either point to use it's x- and y-values (use both from the same point)

x = **2**

y = **3**

Subtract the y-values and x-values to find the slope:

m = $\frac{1}{2}$

Now use basic algebra to solve for b. This is usually the last variable you need to find.

b = **2**

solution:

$y = \frac{1}{2} x + 2$

For the equation of the line you need to have "x" and "y" as variables and "m" and "b" as numbers ("constants")

© Peter Wise, 2015

1. Find the equation of the line going through the points (-1,6) and (-2,-9)

Pick either point to use it's x- and y-values (use both from the same point)

$x =$ **-1**

$y =$ **6**

Subtract the y-values and x-values to find the slope:

$m =$ **15**

Now use basic algebra to solve for b. This is usually the last variable you need to find.

$b =$ **21**

solution:

$y =$ **15** $x +$ **21**

For the equation of the line you need to have "x" and "y" as variables and "m" and "b" as numbers ("constants")

2. Find the equation of the line going through the points (-4,-5) and (8,4)

$x =$ **8**

$y =$ **4**

$m =$ **$\frac{3}{4}$**

$b =$ **-2**

solution:

$y =$ **$\frac{3}{4}$** $x -$ **2**

3. Find the equation of the line going through the points (7,-7) and (-7,1)

$x =$ **-7**

$y =$ **1**

$m =$ **$-\frac{4}{7}$**

$b =$ **-3**

solution:

$y =$ **$-\frac{4}{7}$** $x -$ **3**

© Peter Wise, 2015

61

---

1. Find the equation of the line going through the points (3,4) and (-6,-2).

$x =$ **3**

$y =$ **4**

$m =$ **$\frac{2}{3}$**

$b =$ **2**

solution:

$y =$ **$\frac{2}{3}$** $x +$ **2**

2. Find the equation of the line going through the points (-3,-8) and (4,6).

$x =$ **4**

$y =$ **6**

$m =$ **2**

$b =$ **-2**

solution:

$y = 2(x) - 2$

3. Find the equation of the line going through the points (-9,-2) and (9,-12).

$x =$ **-9**

$y =$ **-2**

$m =$ **$-\frac{5}{9}$**

$b =$ **-7**

solution:

$y = -\frac{5}{9}(x) - 7$

© Peter Wise, 2015

62

TIME	DISTANCE
x value	y value
1	5
2	10

How to figure out slope from a chart

1. Treat the values like x and y values from a table
2. The rate of change is just like finding slope from two points
3. Subtract y's (numerator) and x's (denominator)
4. Your result is the slope!

REMEMBER! ONLY Y-VALUES GO HERE ON TOP!

THESE "RATE OF CHANGE" PROBLEMS ARE JUST LIKE FINDING SLOPE FROM TWO POINTS!

$$\frac{5 - 10}{1 - 2} = \frac{-5}{-1}$$

$SLOPE = \frac{5}{1}$ (rate of change)

negative signs cancel

SLOPE = (rate of change)

1.

TIME (HRS)	MONEY ($)
x value	y value
3	45
4	60

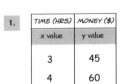

$$\frac{60 - 45}{4 - 3} = \frac{15}{1} \text{ dollars per hour}$$

2.

TIME (SEC)	HEIGHT (FT)
x value	y value
5	200
7	280

$$\frac{280 - 200}{7 - 5} = \frac{80}{2} = \frac{40}{1} \text{ feet per sec}$$

(simplify)

© Peter Wise, 2015

63

1.

Distance traveled in miles

700, 600, 500, 400, 300, 200, 100

Days: 1 2 3 4 5 6 7 8 9

Find the average number of miles traveled from day 2 to day 6

THE ORDER IN WHICH YOU SUBTRACT DOESN'T MATTER! JUST MAKE SURE IT'S CONSISTENT!

#1 Find the points on the lines for day 2 and day 6 (circle these points)

#2 Find the distance traveled for days 2 and 6

#3 Subtract these values and put your answer as the numerator (just like the "change in y")

#4 Now subtract the years and put your answer as the denominator (just like the "change in x")

Subtract the vertical values (like the y-values in regular graphs):

Subtract the horizontal values (like the y-values in regular graphs):

600mi-200mi

$$\frac{\text{Difference in miles traveled}}{\text{Difference in days}} = \frac{400 \text{ miles}}{4 \text{ days}} = 100 \text{ miles per day}$$

6 days - 2 days

$400 \div 4 = 100$

2.

Balance in bank account

$100, $90, $80, $70, $60, $50, $40, $30, $20, $10

Month: January February March April May June July August September October

Find the average change per month in this person's bank account from May to August

$100-$40 = $60

$$\frac{\text{Difference in dollar amounts}}{\text{Difference in months}} = \frac{60 \text{ dollars}}{3 \text{ months}}$$

3 mos.

$= 20 \text{ dollars per month}$

© Peter Wise, 2015

64

1.

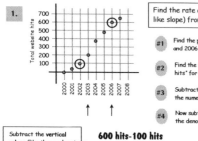

Find the rate of change (just like slope) from 2002 and 2006

#1 Find the points on the lines for 2002 and 2006 (circle these years)

#2 Find the values for "Total website hits" for years 2002 and 2006

#3 Subtract these values and put your answer as the numerator (just like the "change in y")

#4 Now subtract the years and put your answer as the denominator (just like the "change in x")

THE ORDER IN WHICH YOU SUBTRACT DOESN'T MATTER! JUST MAKE SURE IT'S CONSISTENT!

Subtract the **vertical** values (like the y-values in regular graphs):

Subtract the **horizontal** values (like the y-values in regular graphs):

600 hits-100 hits

$$\frac{\text{Difference in website hits}}{\text{Difference in years}} = \frac{500 \text{ hits}}{4 \text{ years}} = 125 \text{ hits per year}$$

2.

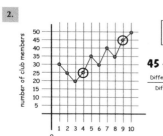

Find the rate of change (slope) between week 4 and week 9

45 - 25 = 20

$$\frac{\text{Difference in membership}}{\text{Difference in weeks}} = \frac{20 \text{ members}}{5 \text{ weeks}}$$

$$= 4 \text{ members per week}$$

© Peter Wise, 2015

65

---

1. Give the meaning of these components of y = mx + b

m = __slope__

b = __the amount you go__ __up or down on the y-axis__ __before starting your first point on the grid__

2.

y-intercept ☐ **2**   Slope = $\frac{1}{3}$

4. Find the y-intercept

PLUG IN EVERY PIECE OF INFORMATION YOU HAVE!

· The slope is 1/4
· The point (8,6) is on the line

y = mx + b   b = **4**

$6 = \frac{1}{4}(8) + b$

6 = 2 + b

3. Make a table and plot the graph

$y = -\frac{1}{5}x + 4$

point values may vary

x	y
0	4
5	3

5. Find the equation of the line with a slope of 2/7 and going through the point (14,12)

$12 = \frac{2}{7}(14) + b$

HINT THINK ABOUT THE PIECE YOU ARE MISSING!

12 = 4 + b   12 = 4 + b

b = 8

$y = \frac{2}{7}x + 8$

© Peter Wise, 2015

66

1.

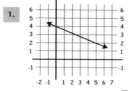

y-intercept ☐ **4**   Slope = $\frac{-2}{5}$

Give the equation of the line:

$y = \boxed{\frac{-2}{5}} x \boxed{-} \boxed{1}$

2. Make a table and plot the graph

$y = \frac{3}{4}x - 1$

USE MULTIPLES OF 4 TO CANCEL OUT THE 4 IN THE DENOMINATOR!

x	y
4	2
8	5
12	8

3. Find the y-intercept

· The slope is 1/3
· The point (9,1) is on the line

y = mx + b   b = **-2**

x = 9 $1 = \frac{1}{3}(9) + b$
y = 1
$m = \frac{1}{3}$ 1 = 3 + b

4. Find the value of the x-coordinate if
· the slope is 4
· the y-intercept is 3
· one point is (x,11)   **x = 2**

y = 11   11 = 4(x) + 3
m = 4
b = 3   8 = 4(x)

5. Find the slope of the line going through these two points:

(2,6) (7,5)

$$\frac{6 - 5}{2 - 7} = -\frac{1}{5}$$

6. Find the equation of the line with a slope of 2/3 and going through the point (9,11)

x = 9 $11 = \frac{2}{3}(9) + b$
y = 11   11 = 3 + b   b = 5
$m = \frac{2}{3}$

$y = \frac{2}{3}(x) + 5$

© Peter Wise, 2015

67

---

1. Slope is defined as "change in ☐ **y over change in x**

2. y = 2x + b has how many solution pairs for x and y?   **infinite**

3. If you have a y-intercept of 4, the x-value of that point is   **0**

4. If you have an x-intercept of 2, the y-value of that point is   **0**

5. When you have an equation like x = 3 the line is ⟨vertical⟩ horizontal (circle one)

Rule with missing x or y values: **x-value is horizontal, but if your equation just has x, the line is vertical; reverse for the y-value**

6. When you calculate slope from two points with $y_2 - y_1$, how do you know which y is y_2 and which y is y_1?

It doesn't matter which one you pick to be y_2 or y_1. You just have to subtract the same way with the x-values.

7. If you have two points: (5,7) and (10,10), find each of the following:

(a) slope   **slope (m) = $\frac{3}{5}$**

(b) y-intercept

(c) the equation of the line   **y-intercept (b) = 4**

Equation of the line:

$y = \frac{3}{5}(x) + 4$

© Peter Wise, 2015

68

Linear Equations Quiz

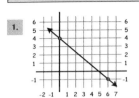

1.

y-intercept [4]   Slope = $-\dfrac{5}{6}$

Give the equation of the line:

$y = \boxed{-\dfrac{5}{6}} x \boxed{+} \boxed{4}$

2. Make a table and plot the graph

$y = \dfrac{2}{5}x - 1$

x	y
0	-1
5	1

3. Find the y-intercept
- The slope is 2/3
- The point (12,1) is on the line

$x = 12$
$y = 1$
$m = \dfrac{2}{3}$

$1 = \dfrac{2}{3}(12) + b$ b = $\boxed{-7}$

$1 = 8 + b$

4. Find the value of the x-coordinate if
- the slope is 3
- the y-intercept is 6
- one point is (x,-6)

$y = -6$ $x = \boxed{-4}$
$m = 3$
$b = 6$

$-6 = 3(x) + 6$
$-12 = 3(x)$

5. Find the slope of the line going through these two points:

$(3,7) \ (5,10) \ \dfrac{\boxed{10} - \boxed{7}}{\boxed{5} - \boxed{3}} = \boxed{\dfrac{3}{2}}$

6. Find the equation of the line with a slope of 2/5 and going through the point (5,9)

$x = 5$
$y = 9$ $9 = \dfrac{2}{5}(5) + b$ $\boxed{y = \dfrac{2}{5}x + 7}$
$m = \dfrac{2}{5}$ $b = 7$

7. Give the equation for the line that goes through the points (14,5) and (7,7)

$x = 3$
$y = 7$ $5 = -\dfrac{2}{7}(14) + b$ $\boxed{y = -\dfrac{2}{7}(x) + 9}$
$m = -\dfrac{2}{7}$ $5 = -4 + b$ $b = 9$

© Peter Wise, 2015

69

Standard Form Equations

Standard Form Equations:

$\boxed{Ax + By = C}$ A, B, and C are constants (numbers as opposed to letters)

Standard form equations have x and y on the left side of the equation

Convert the Standard Form equation to Slope-Intercept Form (y = mx + b)

Standard Form Equation

A. $6x + 3y = 12$ ← Both variables are on the left side

$6x + 3y = 12$
$3y = -6x + 12$ ← Subtract 6x from both sides
$\boxed{y = -2x + 4}$ ← Divide every term by 3

Slope-Intercept Form Equation

Convert the Standard Form equations to Slope-Intercept (y = mx + b) equations

1.
$-2x + y = 5$
$+2x \qquad +2x$ ← Move the term with x to the other side by adding 2x to both sides

$y = \boxed{2x - 5}$ ← Put the term with x first
Slope-Intercept Form

2.
$-7x + y = 14$
$+7x \qquad +7x$ ← Get the x-term to the right side of the equation

$y = \boxed{7x + 14}$

3.
$-11x + 2y = 9$

$2y = 11x + 9$ ← Get the x-term on the right side

$y = \boxed{\dfrac{11}{2}x + \dfrac{9}{2}}$ ← Divide both sides by 2; some terms will be improper fractions
Slope-Intercept Form

4.
$-\dfrac{1}{2}x + \dfrac{5}{3}y = 20$

$\dfrac{5}{3}y = \dfrac{1}{2}x + 20$ ← Move the -1/2x to the other side by adding -1/2x to both sides

$y = \boxed{\dfrac{3}{10}x + 12}$ ← Get rid of the fraction coefficient of y by multiplying every term by 3/5

70

Intro to Point-Slope Form

Point-Slope form is just another way of writing y = mx + b

NOTICE: THE 'M' IS SANDWICHED BETWEEN THE Y'S ON THE LEFT AND THE X'S ON THE RIGHT!

IT'S A SLOPE SANDWICH!

different y-value   different x-

$(y - \boxed{\ }) = m(x - \boxed{\ })$ DISTRIBUTE ON THE SIDE WITH THE X'S!

difference of y's = slope(difference of x's)

Point-Slope Form

Leave these are VARIABLES

$(y - \boxed{y_1}) = m(x - \boxed{x_1})$

THE FIRST LETTERS STAY LETTERS!

THE SECOND LETTERS ARE NUMBERS (THE X- AND Y-VALUES FROM ANY POINT ON THE LINE)!

Constants (NUMBERS) (get this from the x- and y-values from a point on the line)

Example:

A. Put into Point-Slope form: the point (2,1) with the slope (m) = 3

$(y - \boxed{1}) = \boxed{3}(x - \boxed{2})$

Write POINT-SLOPE Equations using the point and slope you are given

1. Put into Point-Slope form: the point (9,2) with the slope (m) = 6

$(y - \boxed{2}) = \boxed{6}(x - \boxed{9})$

difference of y's = slope(difference of x's)

2. Put into Point-Slope form: the point (1,5) with the slope of -4

copy → $(y - \boxed{5}) = \boxed{-4}(x - \boxed{1})$ ← distribute the right side

$(\boxed{y - 5}) = (\boxed{-4x + 4})$

3. Put into Point-Slope form: the point (7,3) with the slope (m) = 2

$(\boxed{y - 3}) = \boxed{2}(\boxed{x - 7})$ → copy and distribute → $(\boxed{y - 3}) = (\boxed{2x - 14})$

© Peter Wise, 2015

71

More on Point-Slope Form

SLOPE of a line

$\dfrac{(y - y_1)}{(x - x_1)} = m$

Multiply both sides by (x − x₁)

$(x - x_1)\dfrac{(y - y_1)}{(x - x_1)} = m(x - x_1)$

Point-Slope form

$(y - y_1) = m(x - x_1)$

You can rewrite this into Slope-Intercept (y = mx + b) form using basic algebra

Put into POINT-SLOPE form, and then convert to SLOPE-INTERCEPT form

1. Point (4,2); slope of 3 $(y - \boxed{2}) = \boxed{3}(x - \boxed{4})$ ← distribute the right side

$(\boxed{y - 2}) = (\boxed{3x - 12})$

add 2 to both sides: $\boxed{y = 3x - 10}$

NOW YOU HAVE THE EQUATION BACK IN Y = MX + B FORM!

AKA SLOPE-INTERCEPT FORM!

2. Point (6,1); slope of $\dfrac{1}{2}$ $(y - \boxed{1}) = \boxed{\dfrac{1}{2}}(x - \boxed{6})$ ← distribute the right side

$(\boxed{y - 1}) = (\boxed{\dfrac{1}{2}x - 3})$

add 1 to both sides → $\boxed{y = \dfrac{1}{2}x - 2}$

3. Point (2,3); slope of 4 $\boxed{y - 3} = \boxed{4(x - 2)}$ ← distribute

$\boxed{y - 3} = \boxed{4x - 8}$

Get the y-term by itself → $\boxed{y = 4x - 5}$

© Peter Wise, 2015

72

Graphing Point-Slope Form

Convert to SLOPE-INTERCEPT form, then graph

1. $y - (-2) = 2(x - 3)$

$(\boxed{y + 2}) = (\boxed{2x - 6})$

$\boxed{y = 2x - 8}$

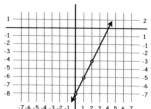

2. Point $(4,-1)$; slope of $-\frac{7}{4}$

$\boxed{y - (-1)} = \boxed{-\frac{7}{4}\ (x - 4)}$

simplify:

$\boxed{y + 1} = \boxed{-\frac{7}{4}x + 7}$

final equation in y = mx + b form:

$\boxed{y = -\frac{7}{4}x + 6}$

3. Point $(-5,-3)$; slope of $-\frac{2}{5}$

$\boxed{y - (-3)} = \boxed{-\frac{2}{5}[x - (-5)]}$

simplify:

$\boxed{y + 3} = \boxed{-\frac{2}{5}x - 2}$

final equation in y = mx + b form:

$\boxed{y = -\frac{2}{5}x - 5}$

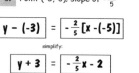

© Peter Wise, 2015

73

Converting Point-Slope to Slope-Intercept

Using basic algebra, you can convert Point-Slope Equations to Slope-Intercept Equations

$$(y - y_1) = m(x - x_1) \qquad y = mx + b$$

A.
$y - 5 = 2(x - 1)$ ← Distribute the 2

$y - 5 = 2x - 2$ ← Add 5 to both sides

$y = 2x + 3$

Convert the Point-Slope equations to Slope-Intercept (y = mx + b) equations

1. $y - 2 = 5(x + 3)$

$\boxed{y - 2 = 5x + 15}$ ← Distribute the 5

$\boxed{y = 5x + 17}$ ← Add 2 to both sides

y = mx + b form

2. $y - 8 = \frac{1}{2}(x - 4)$

$\boxed{y - 8 = \frac{1}{2}x - 2}$ ← Distribute

$\boxed{y = \frac{1}{2}x + 6}$ ← Isolate the y-variable

y = mx + b form

3. $y - 6 = -3(x - 9)$

$\boxed{y - 6 = -3x + 27}$ ← Distribute

$\boxed{y = -3x + 33}$ ← Isolate the y-variable

y = mx + b form

4. $y + 4 = -\frac{4}{5}(x + 5)$

$\boxed{y + 4 = -\frac{4}{5}x - 4}$ ← Distribute

$\boxed{y = -\frac{4}{5}x - 8}$ ← Solve for y

y = mx + b form

5. $y - (-3) = \frac{5}{7}[x - (-2)]$

$\boxed{y + 3 = \frac{5}{7}x + \frac{10}{7}}$ ← Distribute

$\boxed{y = \frac{5}{7}x - \frac{11}{7}}$ ← Solve for y

y = mx + b form $\text{or} -1\frac{4}{7}$

NOTE: ANSWER WILL HAVE FRACTIONS!

SAME WITH THIS CHALLENGE PROBLEM!

6. $y - (-4) = -\frac{2}{5}[x - (-3)]$

$\boxed{y + 4 = -\frac{2}{5}x - \frac{6}{5}}$ ← Distribute

$\boxed{y = \frac{2}{5}x - \frac{6}{5}}$ ← Solve for y

y = mx + b form

© Peter Wise, 2015

74

Review of Standard Form and Point-Slope Form

1. Convert the following equation from Standard Form to SLOPE-INTERCEPT form:

$3x + y = 8 \rightarrow \boxed{y = -3x + 8}$

2. Convert the following equation from Slope-Intercept to STANDARD form:

$y = 7x + 5 \rightarrow \boxed{-7x + y = 8}$

3. Convert the following equation to SLOPE-INTERCEPT form:

$6y = 18x + 12 \rightarrow \boxed{y = 3x + 2}$

4. Convert the following equation to SLOPE-INTERCEPT form:

$-11x + 13y = 15 \rightarrow \boxed{y = \frac{11}{13}x + \frac{15}{13}}$

5. Point $(32,-1)$; slope of $-\frac{3}{8}$

put into point-slope form:

$\boxed{y - (-1)} = \boxed{-\frac{3}{8}(x - 32)}$

simplify:

$\boxed{y + 1} = \boxed{-\frac{3}{8}(x) + 12}$

final equation in y = mx + b form:

$\boxed{y = \frac{3}{8}(x) + 11}$

6. Point $(10,-2)$; slope of $\frac{4}{5}$

put into point-slope form:

$\boxed{y - (-2)} = \boxed{-\frac{4}{5}(x - 10)}$

simplify:

$\boxed{y + 2} = \boxed{-\frac{4}{5}(x) + 8}$

final equation in y = mx + b form:

$\boxed{y = -\frac{4}{5}(x) + 6}$

© Peter Wise, 2015

75

Review: Finding Equations from 2 Points

1. Find the equation of the line going through the points $(1,8)$ and $(3,14)$

x = $\boxed{1}$

y = $\boxed{8}$

m = $\boxed{3}$

b = $\boxed{5}$

slope (m):

$\frac{\textcircled{8} - \textcircled{14}}{\textcircled{1} - \textcircled{3}} = \frac{6}{2} = \boxed{3}$

solution:

$y = \boxed{3}x + \boxed{5}$

2. Find the equation of the line going through the points $(2,-8)$ and $(3,-10)$

x = $\boxed{2}$

y = $\boxed{-8}$

m = $\boxed{-2}$

b = $\boxed{-4}$

solution:

$\boxed{y = -2x - 4}$

3. Find the equation of the line going through the points $(6,3)$ and $(9,5)$

x = $\boxed{6}$

y = $\boxed{3}$

m = $\boxed{\frac{2}{3}}$

b = $\boxed{-1}$

solution:

$\boxed{y = \frac{2}{3}(x) - 1}$

© Peter Wise, 2015

76

Introduction to Linear Inequalities

Inequality basics

1. CHANGE THE INEQUALITY SIGN to an EQUAL SIGN and solve the linear equation normally (y = mx + b)

   a) This line is called the **"boundary"**

2. The kind of line you graph depends on the type of inequality

   a) If the inequality sign is ≤ or ≥ the boundary line will be SOLID

   b) If the inequality sign is < or > the boundary line will be DASHED (to show that "y" DOESN'T INCLUDE the boundary line— y is GREATER THAN or LESS THAN this boundary line)

3. If "y" is **GREATER THAN**, shade **ABOVE** the boundary line; if "y" is **LESS THAN**, shade the **BELOW** the boundary line.

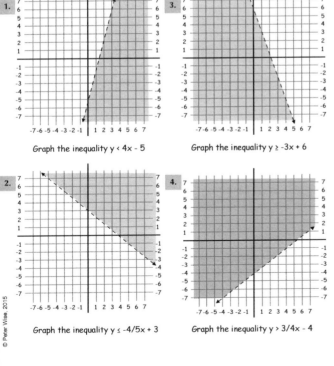

Example:

A.

IF YOU TRY A POINT LIKE (0,2), THE INEQUALITY DOES WORK!

YOU SHADE THIS SIDE OF THE BOUNDARY TO SHOW THAT ALL THE POINTS ON THIS SIDE SATISFY THE ORIGINAL INEQUALITY!

IF Y IS GREATER, SHADE ABOVE THE SOLID OR DASHED LINE!

IF Y IS LESS, SHADE BELOW THE LINE!

"GREATER THAN" here means SHADE ABOVE!

$$y > \frac{1}{2}x + 1$$

- Graph the line as if it were y = 1/2x + 1
- Since y is **greater** (doesn't equal in the original inequality), change the line to a **dashed** boundary
- Since y is GREATER THAN the other part of the inequality, shade ABOVE the boundary line

TESTER POINT TRICK: Plug in any point above or below the line; if the values of a point above the line makes the inequality true shade in that region; if a point below works, shade below.

1. Draw the correct boundary line and shade above or below

$$y > 2x - 1$$

© Peter Wise, 2015

Linear Inequalities, pt. 2

Inequalities with ≥ or ≤

1. Make a boundary line by changing the inequality sign to an equal sign and solving the linear equation normally (y = mx + b)

2. Since ≥ means both > and =, the graph will have a **solid** line (which represents the = part) and be shaded to one side of the line (which represents the > part)

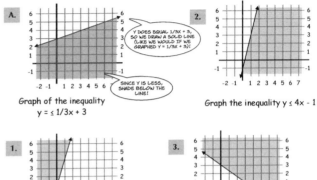

A.

Y DOES EQUAL 1/3X + 3, SO WE DRAW A SOLID LINE (LIKE WE WOULD IF WE GRAPHED y = 1/3x + 3)!

SINCE Y IS LESS, SHADE BELOW THE LINE!

Graph of the inequality
y = ≤ 1/3x + 3

2. Graph the inequality y ≤ 4x - 1

1. Graph the inequality y ≥ 3x + 1

3. Graph the inequality y ≥ -2/3x + 3

Change the ≥ to a = to make a boundary line; if y is greater, shade above the line, if y is less, shade below the line.

© Peter Wise 2015

Linear Inequalities Practice

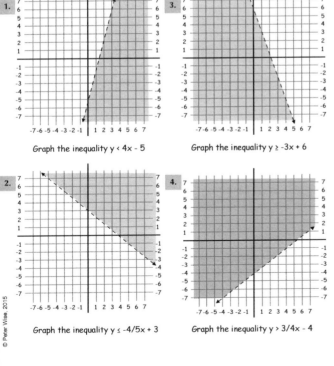

1. Graph the inequality y < 4x - 5

3. Graph the inequality y ≥ -3x + 6

2. Graph the inequality y ≤ -4/5x + 3

4. Graph the inequality y > 3/4x - 4

© Peter Wise, 2015

Review

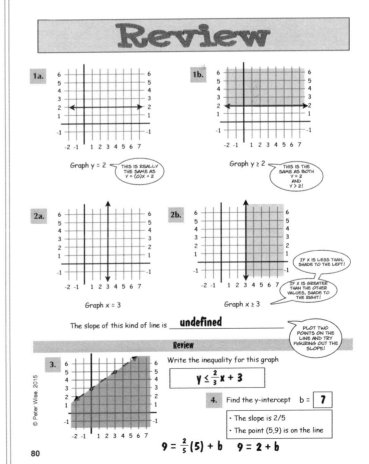

1a. Graph y = 2

THIS IS REALLY THE SAME AS Y = (0)X + 2

1b. Graph y ≥ 2

THIS IS THE SAME AS BOTH Y = 2 AND Y > 2!

2a. Graph x = 3

2b. Graph x ≥ 3

IF X IS LESS THAN, SHADE TO THE LEFT!

IF X IS GREATER THAN THE OTHER VALUES, SHADE TO THE RIGHT!

The slope of this kind of line is ___**undefined**___

PLOT TWO POINTS ON THE LINE AND TRY FIGURING OUT THE SLOPE!

Review

3.

Write the inequality for this graph

$$y \le \frac{2}{3}x + 3$$

4. Find the y-intercept   b = **7**

- The slope is 2/5
- The point (5,9) is on the line

$$9 = \frac{2}{5}(5) + b \quad 9 = 2 + b$$

© Peter Wise, 2015

Review

1. Find the slope of the line going through these two points:

$(3,2)$ $(6,8)$ $\dfrac{\boxed{8} - \boxed{2}}{\boxed{6} - \boxed{3}} = \dfrac{\boxed{6}}{\boxed{3}} = 2$

2. Find the y-intercept if
- the slope is $\frac{4}{5}$
- the point $(5,6)$ is on the line

$y = mx + b$ $b = \boxed{2}$

$6 = \frac{4}{5}(5) + b$

$6 = 4 + b$

3.

Write the inequality for this graph

$\boxed{y < \frac{1}{2}x + 3}$

4. Find the value of the x-coordinate if
- the slope is $\frac{4}{3}$
- the y-intercept is -3
- one point is $(x,5)$

$\boxed{x = 6}$

$5 = \frac{4}{3}x - 3$ $8 = \frac{4}{3}x$

5. Find the equation of the line having a slope of $\frac{3}{7}$ going through the point $(7,12)$

$\boxed{y = \frac{3}{7}x + 9}$

$12 = \frac{3}{7}(7) + b$

$12 = 3 + b$ $b = 9$

6.

Graph the inequality $y \leq -5/3x + 6$

7. Find the equation of the line going through the points $(3,3)$ and $(6,5)$

$\boxed{y = \frac{2}{3}x + 1}$

$x = 3$ $3 = \frac{2}{3}(3) + b$

$y = 3$

$m = \frac{2}{3}$ $3 = 2 + b$

$b = 3$ $b = 1$

Advanced Material: Systems of Equations

- **Solving by Graphing**
- **Solving by Elimination**
- **Solving by Substitution**

Introduction to Systems of Equations

YOUR TASK WITH SYSTEMS OF EQUATIONS: Find values that make all the equations true at the same time

1.

$y = \frac{1}{2}x - 1$

$y = -2x + 4$

Find the point where both lines intersect. The x- and y-values of this point are solutions for BOTH EQUATIONS!

THE X- AND Y-VALUES OF THIS POINT ARE SOLUTIONS FOR BOTH EQUATIONS!

ALL THE X- AND Y-VALUES FOR THE POINTS ON THIS LINE ARE SOLUTIONS OF $y = 1/2x - 1$!

ALL THE X- AND Y-VALUES FOR THE POINTS ON THIS LINE ARE SOLUTIONS OF $y = -2x + 4$!

Prove it

Plug in the x- and y-values of the INTERSECTION POINT into both equations and check to see that they are both valid!

INTERSECTION POINT: $(\mathbf{2}, \mathbf{0})$

$y = \frac{1}{2}x - 1$ $(\mathbf{0}) = \frac{1}{2}(\mathbf{2}) - 1$

y-value of the intersection point   x-value of the intersection point

$y = -2x + 4$ $(\mathbf{0}) = -2(\mathbf{2}) + 4$

2.

$y = x + 2$

$y = 5x - 2$

The INTERSECTION POINT contains the x- and y-values that solve BOTH EQUATIONS

Plug in the x- and y-values of this point into both equations and check to see that they are both valid!

INTERSECTION POINT: $(\mathbf{1}, \mathbf{3})$

$y = x + 2$ $\boxed{3 = 1 + 2}$

$y = 5x - 2$ $\boxed{3 = 5(1) - 2}$

Systems of Equations: Solving by Graphing

Example

A. Find the solution for the following system of equations:

a. $y = x - 3$

b. $y = -4x + 2$

$y = -4x + 2$

ALL THE SOLUTIONS FOR THIS EQUATION ARE ON THIS LINE!

$y = x - 3$

ALL THE SOLUTIONS FOR THIS EQUATION ARE ON THIS LINE!

THE SOLUTION FOR BOTH EQUATIONS IS ONLY THE POINT $(1, -2)$!

STEPS:

#1 Graph the line for each equation

#2 Find the point of intersection

THE SOLUTION TO BOTH EQUATIONS IS THIS INTERSECTION POINT!

Check It Out

PLUG IN THE POINTS $(1, -2)$

a. $y = x - 3$

$(-2) = (1) - 3$

b. $y = -4x + 2$

$(-2) = -4(1) + 2$

DO YOU GET CORRECT ANSWERS FOR BOTH EQUATIONS?

THE POINT OF INTERSECTION IS THE SOLUTION TO A SYSTEM OF EQUATIONS! **Graph each line find where the lines intersect**

1. Find the solution for the following system of equations:

a. Graph $y = 2x + 3$

b. Graph $y = -2x + 7$

c. Find the point where the lines intersect

Solution: $(\mathbf{1}, \mathbf{5})$

= Point of Intersection

= Solution to the system of equations

Systems of Equations: Solving by Graphing

DEFINITION: Finding values that make two or more equations true at the same time

1. Find the solution for the following system of equations:

a. $y = \frac{3}{2}x + 3$

b. $y = -\frac{1}{2}x - 1$

Solution: **(-2, 0)**

= Point of Intersection

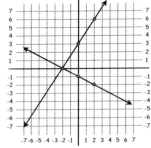

Plug in to check:

a. $y = \frac{3}{2}x + 3$ b. $y = -\frac{1}{2}x - 1$

   $(0) = \frac{3}{2}(-2) + 3$ $(0) = -\frac{1}{2}(-2) - 1$

2. Find the solution for the following system of equations:

a. $y = -2x - 4$

b. $y = 4x + 2$

Answer: **(-1, -2)**

= Point of Intersection

Plug in to check:

a. $y = -2x - 4$ b. $y = 4x + 2$

   $(-2) = -2(-1) - 4$ $(-2) = 4(-1) + 2$

© Peter Wise, 2015

85

Systems of Equations Solving by Graphing

1. Find the solution for the following system of equations:

$y = \frac{2}{3}x + 3$

$y = -\frac{1}{3}x$

WHAT IS THE Y-INTERCEPT IF IT'S MISSING?

Solution: **(-3, 1)**

= Point of Intersection

Plug in to check:

a. $y = \frac{2}{3}x + 3$ $(1) = \frac{2}{3}(-3) + 3$ b. $y = -\frac{1}{3}x$ $(1) = -\frac{1}{3}(-3)$

2. Find the solution for the following system of equations:

$y = 2x - 3$

$y = -\frac{1}{2}x + 7$

Solution: **(4, 5)**

= Point of Intersection

Plug in to check:

a. $y = 2x - 3$ $(5) = 2(4) - 3$ b. $y = -\frac{1}{2}x + 7$ $(5) = -\frac{1}{2}(4) + 7$

© Peter Wise, 2015

86

Systems of Equations Solving by Elimination

EQUATIONS CAN BE ADDED

A.
$+\ \begin{array}{l} x + 3 = 10 \\ x + 3 = 10 \end{array}$

$2x + 6 = 20$

THIS IS REALLY THE SAME AS MULTIPLYING BOTH SIDES OF AN EQUATION BY TWO!

How to Figure Out One Variable at a Time

ELIMINATION METHOD

The y's cancel out so now you can figure out the value of x!

B.
$+\ \begin{array}{l} x + y = 10 \\ x - y = 6 \end{array}$

$2x \quad = 16$

$x \quad = 8$

$8 + y = 10$

$y = 2$

$x = 8 \quad y = 2$

Steps:

#1 Add the two equations to cancel out one variable and to figure out the the other variable

#2 Plug in the value of the variable you know to figure out the value of the one you don't know

Now you know both variables!

SOLVE THE FOLLOWING SYSTEMS OF EQUATIONS USING THE ELIMINATION METHOD

1.
$+\ \begin{array}{l} x + y = 8 \\ x - y = 4 \end{array}$

LOOK FOR THE EASIEST VARIABLE TO ELIMINATE!

2x = 12   #1 Add the two equations to cancel out one variable and to figure out the the other variable

(6) + y = 8   #2 Plug in the value of the x into either equation to figure out y

$x = 6$ $y = 2$

LOOK FOR THE SIMPLEST EQUATION!

2.
$+\ \begin{array}{l} 3x + 2y = 8 \\ 3x - 2y = 4 \end{array}$

6x = 12   #1 Add the two equations to cancel out one variable and to figure out the the other variable

3(2) + 2y = 12   #2 Plug in the value of the x into either equation to figure out y

$x = 2$ $y = 3$

$6 + 2y = 12$ $2y = 6$ $y = 3$

© Peter Wise, 2015

87

Systems of Equations: Solving by Elimination

Solve the following systems of equations using the ELIMINATION method

1.
$+\ \begin{array}{l} 10x + 4y = -12 \\ -10x - 5y = 10 \end{array}$

-1y = -2   #1 Add the two equations to cancel out one variable and to figure out the the other variable

10x + 4(2) = -12   #2 Plug in the value of the y to figure out x

$x = -2$ $y = 2$

2.
$+\ \begin{array}{l} 3x + 5y = 23 \\ -2x - 5y = -22 \end{array}$

x = 1   #1 Add and solve for x

3(1) + 5y = 23   #2 Plug in the value of the x to figure out y

$x = 1$ $y = 4$

3.
$+\ \begin{array}{l} -7x + 3y = 13 \\ 8x - 3y = -14 \end{array}$

x = -1   #1 Solve for x

-7(-1) + 3y = 13   #2 Solve for y

$x = -1$ $y = 2$

4.
$+\ \begin{array}{l} -5x + 3y = 3 \\ 5x + 6y = 6 \end{array}$

9y = 9   #1 Solve for y

5x + 6(1) = 6   #2 Solve for x

$x = 0$ $y = 1$

© Peter Wise, 2015

88

Solve the following systems of equations using the SUBSTITUTION method

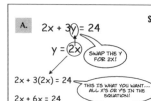

A. $2x + 3y = 24$

$y = 2x$ SWAP THE Y FOR 2X!

$2x + 3(2x) = 24$ THIS IS WHAT YOU WANT... ALL X'S OR Y'S IN THE EQUATION!

$2x + 6x = 24$

$8x = 24$ $x = 3$

Steps:

#1 Look for x = some value or y = some value

#2 Plug in the value of the x or y into either equation to figure out the other letter

#3 Once you figure out one letter, plug in that value into either equation to figure out the other letter

BUT WHAT ABOUT Y?
Plug in x in either equation and solve for y!

$y = 2x$
$y = 2(3)$
$y = 6$

1. $4x + 2y = 40$

$y = 8x$

SOLVE FOR X:

$4x + 2(8x) = 40$ ← Substitute 8x for y in the equation above. Write the (8x) in parentheses

$4x + 16x = 40$ ← Multiply. Copy every other term in the equation.

$20x = 40$ ← Combine terms. Divide by the coefficient of x.

$x = 2$

SOLVE FOR Y:

$y = (8)2$ ← Plug in your value for x into either equation. The second equation is simpler, so let's use that one. Don't forget to put the value for x in parentheses.

$y = 16$ ← Point where the two lines cross (solution to both equations):

$(2, 16)$

© Peter Wise, 2015

89

---

1. $2x + 3y = 4$

$y = 2x + 4$

$2x + 3(2x + 4) = 4$ ← Substitute (2x + 4) for y in the equation above. Write this value in parentheses.

$2x + 6x + 12 = 4$ ← Distribute and rewrite the equation

$8x = -8$ ← Combine like terms (with both variables and constants)

$x = -1$ ← Divide by the coefficient of x

SOLVE FOR Y:

THE POINT WHERE THE TWO LINES CROSS IS THE X- AND Y-VALUE THAT MAKES BOTH EQUATIONS TRUE!

$y = 2(-1) + 4$ ← Plug in your value for x into either equation. The second equation is simpler, so let's use that one. Put the value for x in parentheses.

$y = 2$ ← Point where the two lines cross: $(-1, 2)$

2. $-4x - 2y = 8$ $y = 3x + 16$

SOLVE FOR X:

$-4x - 2(3x + 16) = 8$ ← Substitute the value of y on the right into the equation on the left. Write this value in parentheses.

$-4x - 6x - 32 = 8$ ← Distribute and rewrite the equation

$-10x = 40$ ← Combine like terms (with both variables and constants)

$x = -4$ ← Divide by the coefficient of x

SOLVE FOR Y:

$y = 3(-4) + 16$ ← Plug in your value for x into either equation. Put the value for x in parentheses.

$y = 4$ ← Point where the two lines cross: $(-4, 4)$

© Peter Wise, 2015

90

You can use either equation as long as you get x = some term(s) or y = some term(s)

1. $x + y = 3$ $4x + 2y = 24$

This equation is simpler, so start with this one. → $x + y = 3$ $y = 3 - x$ ← Get the y by itself on the left side by subtracting x from both sides.

YOU COULD HAVE ALSO GOTTEN THE 'X' ON BY ITSELF BY SUBTRACTING 'Y' FROM BOTH SIDES!

YOU WANT TO HAVE 'X' OR 'Y' BY ISELF ON ONE SIDE—SO THAT YOU CAN USE THE OTHER SIDE FOR SUBSTITUTION!

SOLVE FOR X:

$4x + 2(3 - x) = 24$ ← Substitute the terms in the box above for y in the equation.. Write this value in parentheses.

$4x + 6 - 2x = 24$ ← Distribute and rewrite the equation

$2x + 6 = 24$ ← Combine x-terms

$2x = 18$ ← Subtract constants to get them all on the right side of the equation

$x = 9$ ← Divide by the coefficient of x

You could also have used the other equation to get a value for x or y:

$4x + 2y = 24$

$2y = 24 - 4x$ ← Subtract 4x from both sides

$y = 12 - 2x$ ← Divide both sides by 2

SOLVE FOR Y:

$9 + y = 3$ ← Plug in your value for x into either equation. The first equation is simpler, so use that one.

$y = -6$ ← Point where the two lines cross: $(9, -6)$

Challenge:
Plug the values for x and y back into both equations and see if both equations are true →

$4x + 2y = 24$ $x + y = 3$
$4(\ ) + 2(\ ) = 24$ $(\ ) + (\ ) = 3$

© Peter Wise, 2015

91

Start with this equation

1. $2x - y = 4$ $5x - 5y = -10$

$x - y = -2$ ← Divide all three terms by a number to get rid of the coefficient of x.

$x = -2 + y$ ← Add y to both sides. This will leave x by itself on the left side.

SOLVE FOR Y:

$2(-2 + y) - y = 4$ ← Substitute the terms in the box above for x in the first equation.. Write this value in parentheses.

$-4 + 2y - y = 4$ ← Distribute and rewrite the equation

$-4 + y = 4$ ← Combine y-terms

$y = 8$ ← Add constants to get them all on the right side of the equation

You could also have used the other equation to get a value for x or y:

$2x - y = 24$

$-y = 24 - 2x$ ← Subtract 2x from both sides

$y = -24 + 2x$ ← Multiply both sides by -1

YOU COULD ALSO DIVIDE BOTH SIDES BY -1!

$y = 8$ ← Divide by the coefficient of y

SOLVE FOR X:

$2x - (8) = 4$ ← Plug in your value for x into either equation. Use either equation to solve for x. Use parentheses for the value of y that you are substituting

$2x = 12$

$x = 6$

x- and y-values that make both equations true: $(6, 8)$

92

Systems of Equations: Solving by Substitution

Solve the following systems of equations using the SUBSTITUTION method

1. $3x - 3y = 12$
   $-2x + y = 4$

   SOLVE FOR X:

   $y = 2x + 4$

   $3x - 3(2x + 4) = 12$

   $3x - 6x - 12 = 12$

   $-3x = 24$

   $x = \boxed{}$

   SOLVE FOR Y:

   $-2(-8) + y = 4$

   $16 + y = 4$

   $y = -12$

   $y = \boxed{}$

   x- and y-values that make both equations true: (,)

SOLVE THIS ONE ON YOUR OWN!

2. $x - 4y = -5$
   $4x - 2y = 8$

   $x = 4y - 5$

   $4(4y - 5) - 2y = 8$

   $16y - 20 - 2y = 8$

   $14y = 28$

   $y = 2$

   $x - 4(2) = -5$

   $x - 8 = -5$

   $x = 3$

   x- and y-values that make both equations true: (,)

93

94

Slopes Less Than & Greater Than One

This page is information only

A.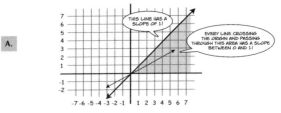

B.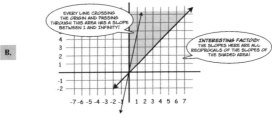

This is a visual proof that there are as many numbers between 0 and 1 as there are between 1 and infinity

C.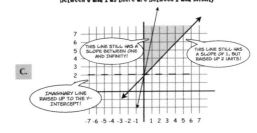

63760342R00071

Made in the USA
Lexington, KY
17 May 2017